Imagining Security

Imagining Security

Jennifer Wood

and

Clifford Shearing

WILLAN
PUBLISHING

Published by

Willan Publishing
Culmcott House
Mill Street, Uffculme
Cullompton, Devon
EX15 3AT, UK
Tel: +44(0)1884 840337
Fax: +44(0)1884 840251
e-mail: info@willanpublishing.co.uk
website: www.willanpublishing.co.uk

Published simultaneously in the USA and Canada by

Willan Publishing
c/o ISBS, 920 NE 58th Ave, Suite 300,
Portland, Oregon 97213-3786, USA
Tel: +001(0)503 287 3093
Fax: +001(0)503 280 8832
e-mail: info@isbs.com
website: www.isbs.com

Hardback
ISBN-13: 978 1 84392 075 5
ISBN-10: 1 84392 075 1

Paperback
ISBN-13: 978 1 84392 074 8
ISBN-10: 1 84392 074 3

British Library Cataloguing-in-Publication Data

A catalogue record for this book is available from the British Library

Project managed by Deer Park Productions, Tavistock, Devon
Typeset by GCS, Leighton Buzzard, Bedfordshire, LU7 1AR
Printed and bound by T.J. International Ltd, Padstow, Cornwall

Contents

Acknowledgements

It is a pleasant experience indeed to be able to look back after completing a piece of work on the many people that have inspired and supported us. Our gratitude is deep. Although we have never felt that we have ever been able to fully realize the possibilities others have inspired, it has been rewarding to try.

We are fortunate in having been able to spend our scholarly lives among so many extraordinary colleagues. This has certainly been true for the time we have spent at the Regulatory Institutions Network at the Australian National University. Some of these colleagues are located here with us physically in Canberra, while others, with whom we continue to maintain very close ties, are scattered widely across the globe. RegNet, in its ANU embodiment and as a wider network, is full of dedicated and talented scholars on whose shoulders we have attempted to climb to see a bit more clearly.

There are people both near and far who have helped us as we have sought to 'imagine security' through this volume. We would like to single out particularly three people. We are enormously grateful to our colleague Nina Leijon who has worked closely with us in putting this volume together. We are also indebted to Julie Grimaldi, our friend and colleague from the Ontario Provincial Police, with whom we have worked closely for nearly a decade. Thank you, Julie, for your astute editorial comments and for verifying the accuracy of what we've had to say about developments in Ontario policing. Finally, a big thank you to Monique Marks, colleague extraordinaire who has done so much to deepen our appreciation of the matters we canvass in this book and who has contributed so enormously to our personal lives. There are, of course, many, many others both at RegNet and elsewhere (many of whom are explicitly recognized in this volume) who have challenged and engaged us. We would also like to acknowledge the wonderful team of people from Victoria Police in Australia who currently work with us on our 'Nexus Policing' Project supported by a grant from the Australian Research Council (LP0348682). To our police team and others with whom we are currently working, thank you for both your criticism and your support in helping us to see more clearly than would have been the case without your insights.

As we write these acknowledgements we, as co-authors, are no longer physically adjacent colleagues. As we move into the role of more 'virtual' colleagues, we would also like to do something a little unusual and recognize the friendship and scholarly support that each of us (with the support of those dearest to us) has accorded the other over many years.

Introduction

How should security be governed at the beginning of the twenty-first century? This question is being voiced with increasing urgency as a variety of new threats to security are being recognized. Our respective levels of fear and anxiety – as human beings, citizens and members of the planet – are accompanied by a growing awareness that the promotion of security – as both an objective and a subjective condition – will require new ways of thinking and acting. This awareness is cementing an important intellectual shift that has been gathering momentum.

Until quite recently, practitioners and scholars have believed – by and large – that the governance of security had evolved to a point where no fundamental change would be necessary for the foreseeable future. These views existed within the context of well-entrenched ideas about governance as centred on the exercise of state capacity in accordance with state-defined interests and goods. Internal security represented the security of a nation state, the promotion of which was its own responsibility that it met through the collection and dispersal of tax resources. External security was promoted in the same manner through agencies designed to protect the integrity of national borders. This was simply how things were, how they should be and how they would be. These ways of thinking, and the institutional arrangements associated with them, were seen as having been perfected through a whole series of incremental changes. We had come – in Fukuyama's (1992) words – to the 'end of history'. What would be required from now on was a constant fine-tuning to ensure that the institutions that had been developed would continue to perform well.

These beliefs stubbornly persisted in the face a 'quiet revolution' (Shearing and Stenning 1981) with respect to both the direction and provision of security governance brought about by the widespread and rapid growth of 'private governments' (Macaulay 1986). Despite the increasingly visible and widespread nature of such changes, many scholars and practitioners continued to cling tenaciously to established state-focused conceptual frameworks. Rather than shifting points of view they have worked harder and harder to fit these phenomena into taken for granted notions of governance authorities and providers (Shearing 2006).

This intellectual complacency was dramatically shattered by the events in New York on 11 September 2001. The US Congressional 9/11 Committee (National Commission on the Terrorist Attacks upon the United States) investigating these events noted how they have changed the way we see the world. Existing ways of thinking and the institutional arrangements associated with them were the deposits, they argued, of an older world. The attacks, they wrote, 'showed, emphatically, that ways of doing business rooted in a different era are just not good enough' (National Commission on the Terrorist Attacks upon the United States 2004: 399). In asking what this meant for US citizens, the Commission argued that Americans 'should not settle for incremental, ad hoc adjustments to a system designed generations ago for a world that no longer exists' (2004: 399).

What the 9/11 Committee did not make as clear as they might have is just how much security governance had already changed underneath, around and alongside established institutions and practices. What 9/11 has done is to bring this already existent world to centre stage. This new worldview has required a rethinking of governance strategies for the future. Security is now being explicitly rather than implicitly re-imagined, as is evident in new kinds of conceptual language. In examining the nature of governance post-9/11, Gross Stein argues that nations cannot govern terrorism effectively without getting their language right. She stresses in particular the importance of 'networks' and 'nodes' as conceptual pillars:

> We are in a new kind of struggle, one against a network with global reach. We need to understand who organizes and manages this particular network. And, we need appropriate conceptual language to understand what a network is, how it operates, how it thrives, and how it withers. Without appropriate conceptual language, we will misunderstand the threat and misconceive the response. (Gross Stein 2001: 73–4)

In imagining a 'network' as a 'collection of connected points or nodes, generally designed to be resilient through redundancy' (2001: 74), Gross Stein goes on to argue: 'No single approach against a single site – even the headquarters, to the extent they exist – will be effective. The implications are clear; removing a single node, or even several, will not destroy the network' (2001: 74).

It is becoming clearer and clearer that both the forms of governance that Western government institutions regard as legitimate and those that they regard as illegitimate share the same fragmented features. Indeed it is now argued that 'dark networks' (Raab and Milward 2003) are doing better at mobilizing a wide range of resources and locating governance outside of state structures than are legitimate forms of governance. What is required is for legitimate forms of governance to catch up with the developments within dark nodes and their networks. Recent developments in local, national and global security provision involve efforts to achieve a 'whole-of-governance' approach that seeks to better coordinate the knowledge, capacities and resources of government, non-government and corporate sector organizations and groupings. Bayley and Shearing commented nearly a decade ago:

> Modern democratic countries like the United States, Britain, and Canada have reached a watershed in the evolution of their systems of crime control and law enforcement. Future generations will look back on our era as a time when one system of policing ended and another took its place. (Bayley and Shearing 1996: 585)

The 'transnational' character (see Sheptycki 1995, 1998b, 2002) of this watershed is something that scholars have only begun to map out and understand. Loader and Walker write:

> We inhabit a world of multi-level, multi-centre security governance, in which states are joined, criss-crossed and contested by an array of transnational organizations and actors – whether in regional and global governmental bodies, commercial security outfits, or a burgeoning number of non-governmental organizations and social movements that compose transnational civil society. It is a world in which policing has, however haltingly and unevenly, been both stretched across the frontiers of states and tasked with combating what are often overlapping problems of global organized crime and political violence. (Loader and Walker in press)

It is this recognition of 'multiple centres', of 'nodes' that has encouraged us to use – in this book and elsewhere – the notion of the 'governance of security' rather than the term 'policing' with its close historical ties to state police organizations (Johnston and Shearing 2003). In order to govern, security programmers, whoever they might be, need to imagine (either explicitly or implicitly) how this might be done. These imaginings, and the mentalities of which they are a part, need to be translated into practices if they are to become more than thoughts. This translation is mediated through institutional arrangements and technologies that are often expressed through habits. All of this requires capacities and resources without which there can be no governance. In the chapters that follow we explore, as our title suggests, the way in which security is imagined and re-imagined and how these imaginings are, or could be, translated through institutional locations into practices that promote security.

Imagining security

What precisely 'security' is, what it should mean, and what should be done to guarantee it, has always been contested. Criminologists continue to reinforce the distinction between 'objective' and 'subjective' senses of the term, particularly given the impact of fear of crime on collective sensibilities and security-seeking behaviour. Zedner writes, for example:

> Security is both a state of being and a means to that end. As a *state of being*, security suggests two quite distinct objective and subjective conditions. And as an *objective condition*, it takes a number of possible forms. First, it is the condition of being without threat: the hypothetical state of absolute security. Secondly, it is defined by the neutralization of threats: the state of 'being protected from'. Thirdly, it is a form of avoidance or non-exposure to danger ... As a *subjective condition*, security again suggests both the positive condition of feeling safe, and freedom from anxiety or apprehension defined negatively by reference to insecurity. (Zedner 2003: 155)

As an 'objective' condition, Zedner points out that crime is 'an ill which, whilst grave, is by no means self-evidently the most damaging source of social misery' (2003: 155). In local governance, notions of 'disorder' and 'incivility' have become central to conceptions of insecurity, leading to new kinds of governing practices. Throughout contemporary developments in public policing, a predominant notion of 'community' functions as the central 'referent object' (Buzan *et al.*1998), in whose name particular security practices and arrangements are justified. Indeed, some see 'community' as replacing 'society' as the main 'spatialization' of governance (Rose 1996).

In the political arena generally, references to *in*security are almost always based on references to crime and, increasingly, disorder. In other words, it is crime that threatens our objective and subjective 'states of being secure'. Criminologists reinforce this near equation between insecurity and crime. If only we can understand criminal behaviour and its precursors, we would live in a much more 'secure' society. To govern security, therefore, is – to borrow from Simon – to 'govern through crime' (Simon 1997). He explains: 'We govern through crime to the extent to which crime and punishment become the occasions and the institutional contexts in which we undertake to guide the conduct of others (or even of ourselves)' (1997: 174). *Crime* thus serves as our main category of thought, the central way in which we frame the security problem. It is the governance of security, *through crime*, that preoccupies most of our efforts. A similar point can be made in respect to global governance and the 'war on terror'. The fear of terrorism has played a considerable role in shaping our collective sensibilities and political practices in the security domain.

With new discourses of 'human security', this may be changing. There exists a widespread political project devoted to 'securitizing' (Buzan *et al.* 1998) threats to various aspects of human life that have traditionally been conceptualized within non-security frameworks (e.g. the economy, the environment or health). With the human being as central referent object, security is no longer understood exclusively in terms of external threats to nations in the form of military attacks. The assumption now is that while posing a significant threat to security, military attacks by other states are only one kind of threat that must be removed, neutralized or avoided. Advocates of a human security approach emphasize the importance of accessible nutritious food, medicines and preventative health care. Security, they say, is also dependent on access to credit and employment (Commission on Human Security 2003). Such structural and development issues are themselves contingent on, and constitutive of, new cultures of peace and the prevention of conflict, particularly in the contexts of the 'new wars' (Kaldor 1999).

The advances made in pursuing a global human security agenda have not led to a decline of state security or 'homeland' security discourses. On the contrary, the current 'war on terror' has involved national governments calling for extraordinary forms of authority and coercive capability in the form of new legislation and the injection of new human and material (particularly technological) resources. Whether or not human security and state security agendas can be aligned is a controversial issue, particularly given that human security is itself threatened by war. Such threats come not only in the form of death and injury, but also in the destruction of social and economic infrastructure that is essential to human development and prosperity (Human Security Centre 2005). What is clear is that the securitization process is a human accomplishment. In imagining security, one is engaged in a 'speech act' (Wæver 1995; Buzan *et al.* 1998). '[I]t is the actor who by securitizing an issue – and the audience by accepting the claim – makes it a security issue' (Buzan *et al.* 1998: 34). The field of securitization is thus one characterized by power struggles (see Dupont 2006a) whereby different actors speak in the name of security on the basis of different forms of knowledge that generate assessments of threat and risk (Bigo 2000).

As the title of our book suggests, 'security' is therefore something we imagine, and what we imagine shapes our mentalities and practices of governance. Security, as Valverde puts it, is not 'something we can have more of or less of, because it is not a thing at all' (Valverde 2001: 85). To govern 'in the name of security' (Valverde and Wood 2001) is to frame governance problems as particular kinds of problems that require specific forms of knowledge, capacities and resources that must be called forth into action.

The practice of governing is also shaped profoundly by our imagination. How we act on the problem of security depends on how we see the world and how we think we can shape events in that world.

Imagining governance

The central conceptual pillar in this book is that of 'governance', which we understand simply as intentional activities designed to 'shape the flow of events' (Parker and Braithwaite 2003). One of the things that life requires is the management of our world which includes, but is certainly not limited to, people. This business of managing our world is the task of governing, or governance.

Governing requires the use of strategies as well as actions intended to put these strategies into effect. This typically takes place in and through institutions that can be thought of as ways of relating people and things to give effect to strategies. Defined in this way institutions can be thought of as 'machines' for doing strategies.

Governing institutions are made up of component parts. Many but not all of these parts are people, while other components include the equipment that people use, like computers. Technologies for getting things done, such as forms that people are required to complete, are also components of institutions. So are the resources that allow institutional outcomes to be created. These various parts, when put together as institutions, enact practices. Governance takes place through programmes that articulate with institutions to produce practices. Another way to think about this is to say that governance requires programmers who develop programmes through institutions.

In deploying the notion of the 'governance of security' we refer to actions designed to shape events so as to create 'spaces' in which people can live, work and play. We are using 'space' as a metaphor that we interpret broadly. Spaces may be conventional territorial spaces but they can also be cyberspaces and social spaces such as communities.

The governance of security has for some time been regarded as the primary responsibility, and indeed exclusive responsibility, of state governments. This has not always been the case. From a historical perspective this way of doing is very new indeed – it constitutes no more than a hiccup in history (Radzinowicz 1948, 1956, 1968).

In developing our understanding of governance we use the concept 'mentality' to refer to sets of ideas that are used to make up the world and make claims about how this world operates. In writing in this way we are recognizing that ideas about the world are not simply ideas about a reality. Rather, they constitute a lens through which to see this world. Mentalities enable us to make up a world and understand the world that they have made up. In what follows we develop a nodal conception of governance. We see this as a conception that is particularly appropriate to contemporary governance in which both the way governance is made up and the way it expresses itself has become increasingly multilevelled. While it is possible to look back on our past and argue that governance has always been 'polycentric' (McGinnis 1999b), our recent past has been characterized by conceptions of governance that are distinctly unicentric. To set the stage for our consideration throughout this volume of nodal governance we now briefly outline the architecture of recent developments in how governance has been conceived and practised.

Governance through force

For the past two centuries the dominant conception of how governance should take place, and how it indeed did take place, has been decisively influenced by theorists, such as Thomas Hobbes (1651), who promoted a centralized conception of good governance. Hobbes provided a wonderful visual summary of his ideas in the picture he selected for the frontispiece of *Leviathan*. In this image governance is pictured as being undertaken by as a benign giant, a Leviathan, who straddles the territory he rules. His body is made up of the people who have agreed among themselves to create him as their ruler. It is this agreement that constitutes his legitimacy as a ruler. He carries a sceptre in his left hand as a symbol of this legitimacy. In his right hand he carries a sword that symbolizes the overwhelming force that he can bring to bear to crush resistance to his rule. Max Weber (1946) expresses this vision of good governance in terms of the idea that the essential tool of governance of states is a monopoly over the legitimate use of force.

Contemporary police organizations in democratic societies have been, and for the most part continue to be, conceived of as the bearers of this state monopoly of force in the domestic security arena. This understanding is expressed by Bittner (1970) in his definition of police as constituting a state-sanctioned source of 'non-negotiable force'. Ayres and Braithwaite's (1992) notion of an 'enforcement pyramid' with state coercion at its apex encapsulates this idea. This idea is expressed in popular culture through its representation of the public police as a 'thin blue line' that separates order from anarchy. It is also symbolized through the tools of the police trade – the baton, the handcuff and the gun.

Anarchy within a Hobbesian mentality is a 'state of nature' in which people's unconstrained pursuit of their concerns and objectives creates a perpetual state of conflict – a 'war of all against all'. Order is created when the pursuit of these concerns and objectives is tamed by the constraint of a state that acts in support of public goods. The Hobbesian social contract between every person with every other person to establish a Leviathan has been given expression, within democratic societies, through elections that select governments to pursue public goods on the behalf of all citizens. The Hobbesian dream is one of a central, and legitimate, source of governance that has overwhelming power to establish and maintain peace. In this top-down, 'command and control' vision, governance is conceived as a pyramid, with legitimate force at its pinnacle.

For Latour (1986), this conception of the power to govern is similar to the idea of inertia within the physical world. Movement, from an inertial perspective, is understood as the result of an initial force. The movement, once initiated, continues until it meets resistance. Unless resistance undermines the initial force the movement continues. In this pinball conception, power is 'owned' by its source – others may frustrate it and even mobilize it, but it remains centrally located.

Governing through enrolment

In contrast to the Hobbesian conception, Foucault (1990) conceives of power, and hence governance, as widely shared. Power is understood as being everywhere, not because it is exercised everywhere, but because it is viewed as coming from everywhere. Within this conception power is not something that can be 'owned'. It is not something that one can 'possess and hoard' (Latour 1986: 265). Latour develops this alternative conception by thinking of power 'not according to the power someone has, but to the number of other people who enter into the composition' (1986: 265).

The power to shape events is produced, for Foucault and Latour, not by owning it but by enrolling others to perform actions required to realize one's objectives. Power, and hence governance, is produced through 'action at a distance' (Latour 1987). Within this conception, power is a 'convenient way to *summarise* the consequence of a collective action' (Latour 1986: 265, italics in original).

Both Latour and Foucault insist that we should not conceive of power as a cause of things, as the inertia conception of power maintains, but as an *effect*. To appreciate this, Latour argues that we need to see power not in terms of the idea of 'diffusion' implicit in the inertial metaphor but as a matter of 'translation'. To make this point he uses the metaphor of a rugby game. A rugby game is not determined, he points out, by the first kick. This kick is simply the first of a whole train of events, none of which is caused in any way by this first kick. Every player who holds the ball must make a decision as to what to do with it. This acting together is the cause of any power that the team might have. Governance is thus 'the consequence of an intense activity of enrolling, convincing and enlisting' (Latour 1986: 273).

Based on this dispersed conception what is required for effective governance is the enrolment of a range of persons with a range of knowledges. Put differently, what is required is the enrolment of distant actions in pursuit of governance ends, such as the pursuit of the common good that Hobbes saw as the legitimate objective of governance.

One of the most compelling ideas developed to answer this way of thinking about governance has been that markets – understood not simply as places of exchange of goods but as the source of coordinated enrolment – can be used to promote effective governance. Within this conception the Hobbesian picture of a giant standing astride a country and made up of people retains its meaning, but this giant is now seen not as a person or set of people but as an expression of markets. Smith (1776) imagines an enormous 'invisible hand' that coordinates and organizes diffuse capacities and knowledges as power resources to shape the flow of events.

Free markets, that is markets left to work independently, are thought of in this conception as having the capacity to coordinate the production of public goods (Hobbes' 'commonweal') without the intervention of a Leviathan. Free markets are conceived of as mechanisms that create the 'intense activity of enrolling, convincing and enlisting' that good governance requires within a Latourian understanding. A term that has come to be associated with this idea of governance that works through free markets is 'liberalism' or liberal governance. Within this conception liberty is essential because governance requires liberty – in Rose's (1999) language, one 'governs through freedom'.

A conception of governance that brings together something of both the conceptions we have outlined and that has been, and continues to be, very influential is 'neo-liberalism'. This hybrid conception brings together states as the source of direction for governance and markets as a source of provision. This conception is critical of the classic liberal idea that free markets will on their own act in ways that are congruent with the public interest while retaining the idea that markets work well as sources of coordination. Markets, and specifically the private sector, should provide while governments (in particular democratically elected governments) should direct.

This mentality of governance refigures both the Hobbesian and Smithian conceptions by separating out the 'steering' of governance from the 'rowing' (Osborne and Gaebler 1992). This way of making up the world of governance argues essentially that Smith was right in saying that non-state institutions are best at doing the provision of governance (and that markets are useful vehicles for doing so) but that he was wrong in arguing that markets would operate to ensure that a commonweal was realized. Ensuring that a commonweal is established is something that states rather than markets should do. At the same time, providing for the commonweal is something that markets rather than governments should do.

As this neo-liberal understanding has been developed the idea of 'markets' has emerged as a metaphor for dispersed provision that can be done through a variety of non-state mechanisms. Two terms that have been introduced to elaborate on this conception are 'partnerships' and 'networks'. The idea these terms are used to capture is that governance should take place through state-directed partnerships and networks. State direction within this understanding is seen as ensuring that democratically elected and accountable governments set goals and then work to enrol others through market-like mechanisms to realize desired outcomes. This neo-liberal idea has been succinctly captured by Loader and Walker (2006) as an 'anchored pluralism'. Within anchored pluralism governance is plural, but is at the same time orchestrated by state government in ways that seek to ensure that state-defined objectives are accomplished.

This way of thinking about governance has been used both to understand developments within the governance of security and to shape it. The following chapters question all three of the conceptions we have just reviewed – command and control, classic liberalism and neo-liberalism. We hope to demonstrate that while they do embody useful conceptual elements their accounts are inadequate as either explanatory or normative accounts. Over the course of this book we build an alternative account of governance – a nodal account. We use this both as a descriptive framework and as a frame within which to imagine possibilities for reshaping the governance of security. This nodal account recognizes networks, partnerships, markets and states as sources of governance and locates such sources within a field of organizational nodes. We explore these nodes by considering how their mentalities, practices and resources are, or could be, articulated as locations of capacity and knowledge engaged in shaping the flow of events.

In Chapter 1 we develop more fully our conception of the nodal governance of security and begin exploring the explanatory and normative themes that we pursue throughout this volume.

Chapter 1

From state to nodal governance

Introduction

In the previous section we argued that there have been fundamental changes in the way in which governance has been understood. We began our discussion with an analysis of the established view of security governance as something that should be undertaken by states through the application of legitimate physical force. This is a view of states and governance that is often associated with the work of Weber, who defined states as entities that were able to acquire a legitimate monopoly over the use of physical force (Weber 1946).

These Weberian ideas have a long lineage. A crucial moment in the development of this conception of the state and of state governance was Hobbes' *Leviathan* (1651/1968) published in the middle of the seventeenth century. Another important marker in the history of the nation state was the Treaty of Westphalia 1648, which sought to articulate and cement a state system within Europe. States in terms of the Treaty were to be regarded as the ultimate sources of territorial governance.

This state-centric conception has been reshaped over the past half century by developments in indirect governance that rely on the enrolment of others in performing actions required to realize the objectives of a particular governing authority. This conception is nicely captured in Latour's notion of 'action at a distance' (1987). As an illustration of indirect governance we referred to the ascendancy of neo-liberalism in various parts of the world.

In this chapter we build the argument that neo-liberalism, and the dream of state action at a distance that it represents, is only one part of a much bigger story of 'nodal governance'. We use this term to denote a multiplicity of governance authorities and providers that coexist in multiple ways to produce diverse security outcomes. The story of the nodal governance of security is one of hybridity. Governance is not performed simply by institutions of the state, nor shaped solely by thinking originating from the state sphere. Today, ways of imagining and realizing security governance in the business sector as well as the 'third sector' (e.g. community groupings, non-governmental organizations) shape and influence the thinking of state institutions and vice versa. This is the essence of nodal governance. While this chapter attempts to make the case for a nodal perspective, the following chapters explore dimensions of this hybridity theme. Chapter 4 focuses on the normative implications of our analysis for what Braithwaite would describe as 'weak actors' (Braithwaite 2004) in the security field.

Transformations in state governance

The nineteenth century saw the emergence of a fundamental shift in the way security was thought about and practised across much of the world. Within the English-speaking world the changes promoted by then Home Secretary, Sir Robert Peel, to establish a new 'modern' police institution in London in 1829 – a specialized agency made up of persons who belonged to a specialized occupation – became emblematic of the host of changes that took place to recognize the police as the central provider of security governance. These changes emerged in response to what was perceived by many as a crisis in the delivery of security via a plurality of largely 'private' arrangements with feudal roots and resonances (Shearing 1992).

There has been much debate over what was unsatisfactory about the previous security system, but what is clear is that the new arrangements were thought of as providing more effective and democratic security governance. While there has also been debate about who benefited from these developments, the changes were promoted in the name of a collective or 'public' interest. This link between the public interest and the police is reflected in the term 'public police', as well as in the idea that the police are a 'law enforcement agency' – within this conception the law is understood as expressing a general public will.

The police were thought of as the end product of a historical progression that had culminated in a situation where the state should establish a monopoly over security and other domains of governance. This arrangement was here to stay and would not be challenged in any thinkable future. Within this way of seeing, what was required was to understand, extend and improve upon this police monopoly. As the idea of a state monopoly had won the day, all that was required was simply action to realize this ideal as fully as possible. This involved perfecting the police as an institution.

This state-centric view has remained remarkably resilient in the face of a broader shift from welfarism to neo-liberal thinking. While advocates of welfarism conceptualized states as both authorities and providers of governance, neo-liberal thinkers sought to uncouple the 'authority' and 'provider' functions (Bayley and Shearing 2001), expressed in Osborne and Gaebler's simple yet powerful prescription for states to 'steer' more and 'row' less (Osborne and Gaebler 1992). The devolution of rowing has been justified on the grounds of efficiency and economy in service provision, while the retention of the steering function is seen as central in order that the 'public' good of security is distributed in accordance with the 'public' interest. To steer governance is not only to authorize and legitimate rowing, it is to ensure that this rowing is regulated.

What scholars of the 'new regulatory state' (Braithwaite 2000a) argue is that neo-liberalism has not been accompanied by deregulation, but on the contrary has been accompanied by a growth in regulation by or on behalf of states (Braithwaite 2000a; Levi-Faur 2005). What we hope to illustrate in this chapter, and through this book, is that Osborne and Gaebler's dream for 'reinventing governance' is a simplistic representation of what in practice has been a diversification not only in rowing functions but in steering functions as well. Before moving through the nodal governance story of multiple auspices and providers, we continue with the neo-liberal story and how it is expressed through practical ways in which states govern through others.

Governing through others: enrolment and alignment

Today, states seek to govern indirectly through mobilizing the knowledge, capacity and resources of other institutions, groupings and individuals in the delivery of security and other goods. In some cases, state actors are not endeavouring to rethink the ways in which they imagine security and its governance, but are rather concerned with expanding their capacities and resources in furtherance of their mission. As we discuss in Chapter 2, on 'waves' in public policing, there have been and continue to be explicit attempts on the part of state actors – like the police – to incorporate other mentalities to the end of transforming their own strategies and practices. Nevertheless, there is much that occurs that is aimed largely at bolstering state capacity to 'govern through crime'.

For instance, as part of a broader trend in 'compelled third-party participation in the regulatory process' (Gilboy 1998), the police can command private actors to assist in law enforcement activities (Ayling and Grabosky 2006). In the particular area of 'third-party policing' (Gilboy 1998), various 'regulatory "nodes" (including both willing and unwilling partners) come together to solve a crime problem' (Mazerolle and Ransley 2005: 2). The term 'third-party' is used in reference to those actors who 'are neither the primary authors nor beneficiaries of the misconduct they police' (Gilboy 1998: 135), such as public housing agencies, property owners, parents, health and building inspectors, and business owners, who take some responsibility for preventing crime or reducing crime problems (Mazerolle and Ransley 2005).

The capacity and authority of the police to command third-party participation in crime control rests on the use of 'legal levers' from criminal, civil, regulatory or other laws (Mazerolle and Ransley 2005). Mazerolle and Ransley explain that laws and regulations originally designed for purposes other than crime control, such as Health and Safety Codes, or Uniform Building Standards, often serve as such levers. In fact, they add that the trend towards third-party policing must be understood within the context of an increased 'blurring' of civil and criminal laws whereby the previously distinct categories of law are converging. They write: 'Increasingly, the criminal law makes use of civil processes and remedies, while in both regulatory and private law serious misbehaviours are criminalized' (2005: 67).

Mazerolle and Ransley provide various illustrations of third-party policing efforts that target different 'focal points' of criminal behaviour which they describe as 'people (individuals or groups of people), places or "scenes", situations, crime targets (victims), accomplices, and props (instrumentalities)' (2005: 66). For example, police can coerce a property owner to evict a tenant engaged in criminal or disorderly activities. Police may threaten 'duty of care' civil suits against licensed premises for serving or pricing alcohol in ways that can increase the risks of crime and violence. Those who control the 'props' of crime – those objects or devices that enable the commission of a crime, such as cell phones or cars – can be commanded to assist in crime control through the use of legal levers (2005: 66). Even parents can be enrolled to govern the behaviour of their delinquent children. They can even be prosecuted for activities carried out by their children. In the state of Queensland, Australia, parents of truant children can be fined $375 for a first offence and $750 for further offending behaviour (2005: 57).

Mandatory reporting to state authorities is another manifestation of third-party policing. The classic example is the requirement on banks to report cash transactions over a certain amount (Grabosky 1995: 530). In the area of drug control, US chemical companies involved in production and sale are required to report sales in quantities above a certain threshold to the Drug Enforcement Administration (Cherney *et al.* 2005: 19).

A further example of third-party policing comes from the Crime and Disorder Act in the United Kingdom, which mandates local governments and police to engage in crime and disorder reduction activities and partnerships. In so doing, such partners must undertake analyses of crime and disorder problems and patterns and establish plans that are managed according to short- and long-term performance targets. They must also assess the effects of their various municipal government activities on crime and disorder in the local government area. The Act also provides a set of civil legal levers, such as Anti-social Behaviour Orders, 'directed at any person aged 10 or over acting in a way likely to cause harassment, alarm or distress to one or more other people', parenting orders, 'made for children subject to child safety, sex offender or anti-social behaviour orders, or truancy offences, requiring the child's parent to comply with conditions including counselling or guidance sessions', and child curfew schemes that enable local authorities to 'restrict the time when young people can be out in public places during nighttime [or, in certain jurisdictions, even daytime] hours' (Mazerolle and Ransley 2005: 156).

Third-party policing is initiated by a range of actors including but not limited to the police, prosecutors, government agencies, regulatory agencies, community groups, businesses, and even citizens themselves. For example, airports are required by aviation regulators to adopt particular screening practices designed to reduce instances of illegal immigration or terrorism. A range of professionals including doctors, auditors and psychologists are required to report suspected illegal activities. In such cases, third parties are compelled 'to engage in practices that are potentially outside of their routine activities in an effort to prevent or control crime problems' (Mazerolle and Ransley 2005: 178).

If third parties do not undertake the functions for which they have been enlisted, they could incur severe criminal or civil penalties. This can include fines, loss of licence (for example, in the case of physicians who prescribe drugs for non-medical purposes), disruption of business (for example, through raids of businesses that utilize illegal workers), and confiscation of goods (for example, of those produced by manufacturers that rely on sweatshops) (Gilboy 1998: 142).

In addition to command and control mechanisms for enrolling others in security governance, states employ a variety of other strategies, including the use of incentives or rewards (Grabosky 1995: 534). Rewards are commonly used in encouraging citizens to come forward with information that could contribute to a police investigation (1995: 535). For example, in the United States a Counter Narcotics Reward Program was established in 1986 to assist in the identification and prosecution of narcotics traffickers (Cherney *et al.* 2006).

The use of financial rewards is often not required in 'responsibilizing' (O'Malley and Palmer 1996) others to further state objectives. A common way in which indirect governance has been realized is through mobilizing the knowledge, capacity and resources of volunteers who constitute what Drucker describes as the 'third sector' (Drucker 1994). Some years ago Drucker noted that Americans devote a good deal of their 'free' time outside of work to voluntary activities. He saw this as a tremendous untapped resource that governments could mobilize to get the job of governance done (ibid.). In fact, this idea of mobilizing the resources of volunteers has always been important to the state police who have, for example, used the dispatch system (e.g. 9-1-1 in the United States) to enable victims to call on the police for help and to transmit information to the police. Both the 9-1-1 system and 'Crime Stoppers' programmes transform citizens into informers (Shearing 1984) or what Ericson and Haggerty would describe as 'knowledge workers' (Ericson 1994; Ericson and Haggerty 1997).

Indirect governance is also carried out through the use of market mechanisms or what Crawford depicts as 'contractual governance' (Crawford 2003). State governments not only use tax resources to fund state agencies, but buy governmental services through a market.

Private security firms, for example, provide a wide range of security services under contract to governments, such as the guarding of government buildings, including police buildings, and the running of prisons (Ayling *et al.* in press; Jones and Newburn 2006). These contractual and market processes have had the effect of 'commodifying' governance (Loader 1999). This has meant that security has become a commodity that is bought and sold in a marketplace. What this does by implication is to redefine state agencies, not simply as the providers of governmental services but as contractors for the delivery of 'public' goods.

Marketization has not been limited to the delivery of security outcomes at local or domestic levels. The private provision of military services is undertaken routinely by states through contractual arrangements (Singer 2003; Avant 2005). The following story was told in April 2004 in the Melbourne-based paper *The Age* at the time when four employees of an American firm were killed in Iraq and their bodies dismembered and paraded in the streets:

> When the door opens at Level 5 in the Palestine Hotel, there's a spit-and-polish Gurkha pointing a high-powered gun into the lift.
>
> The whole floor and another above it has been taken by Kellogg Brown & Roots, the construction wing of Halliburton, which is one of the biggest US firms working in Iraq. And although the linguists of occupation don't allow the word 'mercenary', the Gurkha is part of a 15,000-strong sub-contract security operation that is the third biggest armed force serving in Iraq.
>
> Their numbers – and salaries as high as $US1000 ($1310) a day – attest to the danger of this Arab version of Dodge City.
>
> . . .
>
> The ranks of the private armies in Iraq are growing so rapidly that US and British defence officials are at a loss on how to counter offers to the best of their special operations and SAS staff; and the risks are so great that their new employers complain that as much as 40 cents in every dollar goes into an insurer's pocket.
>
> . . .
>
> In the mayhem, Baghdad has been carved into a series of Western security bubbles.

> There is the Green Zone, American proconsul Paul Bremer's sprawling bunker for which the Pentagon is about to let a $US100 million privatised security contract; there is the blast-walled compound of the Palestine and Sheraton hotels, which have their own militia; foreign embassies are grouping and fortifying; and Western business and the foreign media have all but withdrawn behind wire and guns, reluctantly venturing out with security vehicles.
>
> . . .
>
> And it's not just the foreigners – the South Africans who know they are breaking their country's laws on mercenary activity; the skilled Gurkhas and Fijians; or the Chileans who trained during the dictatorship of General Pinochet.
>
> Beneath all that is a layer of Iraqi-run security – hundreds of local firms with the capacity to become private, clan based militias ... (Mcgeough 2004)

This dramatic account provides an example of how security is being governed around the world by the United States and many other countries. In Iraq, the governance of security has been undertaken by the US government through assemblages of capacities drawn from around the world from the public and the private sectors.

In the case of marketization, the notion of 'steering' is very different from that of a top-down command and control conception. Instead, all those enrolled in governance shape it, or to use Latour's term 'translate' it 'according to their different projects' (Latour 1986: 268). Enrolled actors align the direction given by a centre with their own objectives, and in doing so change the directions. The result is that the direction is, and indeed must be, translated into terms that are compatible with others. The art of governance here is the art of alignment (Deukmedjian 2002).

This idea of multiple objectives that need to be aligned is what Foucault alludes to when he writes about governance as the right distribution of things for a 'convenient end' (1991: 93). In this form of governance direction is only possible if a plurality of convenient ends can be brought into sufficient alignment to allow the objectives of governments to be realized. What emerges, to use Rose and Miller's terms, is 'a profusion of shifting alliances' (Rose and Miller 1992: 174). Foucault describes this difference between direct governance and governance at a distance nicely when he relates command and control governance to taking someone by the hand and leading them. In this mode of governance, he claims, one forces the person to go here or there because one has the physical force at one's disposal to permit one to do so. In contrast, with rule-at-a-distance governance one does not force; rather one enrols and persuades. In this situation the person remains free to choose not to go along with one's directions.

To use Rose's term, one 'governs through freedom' (Rose 1999). This form of governance requires free subjects. In the case of the story of Iraq, the Gurkhas or a paramilitary company or other organizations are 'free' to enter into a contract intended to align their objectives with those of the government involved, or with some company that is subcontracting particular services. In each case the person (or persons) chooses to act in ways that are aligned with a particular governing objective because it fits with their objectives (Foucault 1988: 2). Through the marketization of military functions, the US government, in the above example, enables these private actors or companies to make a profit by providing services to them in exchange for money. At the same time, the situations within which such contracts are entered vary considerably so that for some there are more choices than there are for others. Power is unequal.

This contracting out of security provision to non-state providers is an important manifestation of nodal governance. Governments are continuing to constitute markets in ways that allow state-defined interests to be promoted through their service delivery arrangements with commercial providers, as we see in the above example of the Gurkhas. In other words, this is not a case of unfettered free markets for security goods. Furthermore, some state agencies like the police have themselves entered the marketplace to compete with other state and non-state providers in order to secure a central place for themselves in the security delivery field (Blair in press). This may be partly the result of an angst engendered by a growth in commercial providers who are increasingly offering products and services traditionally understood as a 'core' function of the police or the military (Wood 2006b). By deliberately setting themselves up as agencies designed to compete with other state police organizations as well as with private security companies (Wood 2000), the police are serving to reinforce their 'public' character and the normative position that policing remains a public good (Loader and Walker 2001, 2006).

An example of an ambitious effort to constitute markets by actively competing in them can be found in the establishment of 'community support officers' in the London Metropolitan Police.

Sir Ian Blair, the Commissioner of the London Metropolitan Police (the Met), argues that public police organizations need to acknowledge the plural nature of security delivery (Blair 1998). He depicts a 'community safety market' that includes not only private security but also the Neighbourhood Warden or Street Warden schemes (Blair 2003; Johnston 2003). What is distinctive about such schemes, Blair notes, is that they do not operate on privately owned spaces, but rather function through the provision of patrol on public spaces. Blair adds: 'It is certainly fair to state that the British private security industry sees these developments as offering a significant potential market, capable soon of being exploited' (Blair 2003: 1). This 'wake up call' (2003: 1) has prompted Blair to reflect on where police should fit in to this plural landscape, both in terms of their functions and their relationships to other security providers. Marketization, as this example illustrates, is simultaneously promoted and resisted by states.

The reality of nodal governance, that we have been outlining, is much messier than the neo-liberal narrative would suggest. What one has in practice is not a single model of governance, but a complex of hybrid arrangements and practices in which different mentalities of governance as well as very different sets of institutional arrangements coexist. We have not simply witnessed a shift away from direct command and control governance to forms of indirect state governance that operate through market mechanisms or through the gentle touch of persuasion associated with third-sector mobilization. Rather, we have a complex set of relationships in which 'steerers' and 'rowers' constitute relationships and align their interests.

In the remainder of this chapter and throughout this book we seek to demonstrate that it is not simply state auspices, or state steering agencies, that enrol others and align objectives within a rule-at-a-distance framework. Security governance is also authorized and undertaken by a range of private authorities that govern both directly and indirectly according to various mentalities. Indeed, from the point of view of any auspice of governance the world looks like a series of possibilities for enrolling others as providers of governance in realizing their objectives.

We hope to demonstrate that the story of 'private governments' (Macaulay 1986) brings us closer, but not close enough, to the meaning of nodal governance. In the last part of this chapter we suggest that it is not sufficient to simply distinguish between the mentalities, institutions and practices of 'public' and 'private' auspices and providers. Rather, ways of thinking and acting within the public and private spheres shape and influence one another. Nodal governance is thus characterized by institutional hybridity as well as hybridity in ways of thinking and acting.

Private governments

The 'quiet revolution' in 'private policing' that Shearing and Stenning (1981) observed over a quarter of a century ago is not so quiet any more. Whereas the early days of private policing saw developments taking place within the context of privately owned spaces that were accessed and utilized by private parties, this is no longer the case. Indeed, members of the public are arguably governed, within both public and private arenas, just as much by private authorities and providers as they are by public ones.

The huge increase in private policing that took place during the last half of the twentieth century was possible because it conformed with, and indeed was deeply embedded in, well-established provisions of property and contract law (Hermer *et al.* 2005). When commercial forms of security were initially utilized they operated almost exclusively on 'private property' as agents of property owners. This fitted well with the established view that the state police acted to promote security in the public interest while private providers did so to promote private interests.

This neat separation was spoiled with the emergence of 'mass private property' (Shearing and Stenning 1981, 1983a) – shopping malls and theme parks, for instance, that while privately owned function as public spaces. In these spaces corporate providers are employed to provide security on behalf of the owners of these spaces in furtherance of their interests and the interests of those who lease space from them. Thus, along with this blurring of space, came a blurring of the notions of the 'private' and the 'public' interest. Here we had private interests acting to govern 'common' spaces (von Hirsch and Shearing 2000; Hermer *et al.* 2005) in which 'common' interests (Shearing and Wood 2003a), not just 'private' interests, were important. These included the interests of groups such as residents, shopkeepers, consumers and wider society.

Other forms of 'communal governance' have also become widespread. A typical example involves the clubbing together of businesspeople or residents in associations that act to govern security not simply within mass private property but on public streets (Murphy 1997). Under the auspices of 'business improvement districts' and the like, security and other 'public' services are carried out in ways that complement those provided by state agencies. This provision of governance is thus directed by common interests that are aligned with but not identical to broader public concerns and objectives (Shearing and Wood 2003a).

In South African residential improvement districts, residential property owners not only hire private security to protect individual properties through emergency response services and the installation of alarm systems, but also arrange for agents to patrol the public streets on which they live. In both cases what has happened might be thought of as a contemporary form of enclosure. Here non-state auspices now govern space that was once governed exclusively by states (Shearing and Kempa 2000, 2001).

These and similar developments have encouraged and fostered the idea of public-private 'partnerships' in the provision of public services. These partnerships do not simply involve state agencies contracting out provision to private sector organizations as discussed earlier as part of a move to reinvent government. Rather, arrangements for undertaking 'public' services, that are sometimes explicit while at other times implicit, are now shared between a variety of state and non-state agencies that operate under the guidance of different auspices. The negotiations that these arrangements sometimes involve constitute a form of 'democratic bargaining' (Drahos and Braithwaite 2002: 13). An example of such bargaining, to use another South African illustration, is an auspice that runs a large waterfront residential and commercial development. This corporation has a bargained agreement with the City of Cape Town that accords it tax relief for providing public services within this development.

Private property has come to function as a double-edged sword. While it continues to be a source of protection against state intrusion into people's private lives, it has also become the legal basis for the emergence of private governments. In the spaces in which these governments operate, property law legitimates private engagement within common spheres. Indeed, as private property becomes public space and as businesses and residents commission agents to govern these spaces to protect themselves and their property (see Noaks 2000), the public/private distinction loses its established meanings. Public space can no longer be equated with public governance.

The world of commercial security provision further demonstrates the plurality of ways in which both direct and indirect governance is realized. Multiple auspices govern a wide variety of domains, both physical and non-physical that range from gated communities to cyberspace, in furtherance of a wide range of agendas that relate to and align with state agendas in many different ways (Wall in press).

Private governments take into account, and seek to take advantage of, the objectives and regulatory constraints that states have developed. In this they are no different from states themselves who also take into account the objectives and constraints of the regulatory frameworks of other governing entities. The order maintained by a governing entity such as General Motors at its motor plants or by the Disney Corporation at its recreational sites is not simply designed to comply with the governance objectives of states. These corporations have their own objectives that they attempt to put into effect through a variety of governing strategies. In doing so they take into account the objectives of state agencies and may seek to enrol these in assisting them in corporate governance if this seems sensible.

An example of this can be found in the financial regulation of stock exchanges such as those in Toronto and New York. A Canadian study found that these exchanges often deal with the police, but when they do so it is not in their capacity as entities enrolled by the police. Rather, they do so as entities that seek to enrol the police. It is their agendas rather than police agendas that shape their self-governance (Stenning, Shearing *et al.* 1990). Much the same applies to any financial institution when it promotes its corporate orders. As seen in the domain of third-party policing, banks are sometimes enrolled by the police to promote state-defined objectives, but they are much more likely to be engaged in enrolling the police in promoting their objectives and agendas.

As with state action at a distance, indirect governance on the part of private authorities comes in many forms. As Roach Anleu and colleagues point out, private entities also enlist third parties in crime control and do so through various mechanisms including positive sanctions such as financial rewards or other incentives (Roach Anleu *et al.* 2000: 72). A good illustration of 'market-based third-party policing' (2000: 72) can be found in the insurance industry. Property insurance companies undertake activities 'that constrain the activities of actual or potential policy holders, and that are aimed at reducing criminogenic opportunities' (2000: 73). Through financial incentives, insurance companies enrol policy holders in minimizing their risks of victimization while recruiting otherwise 'low-risk' clients in order to spread risks across the client base. Here, the interests of insurance companies are aligned with those of their clients; the former sustain profits while the latter benefit from a competitive premium (2000: 74).

It could be said that enrolment of third parties by public police and other state agencies reflects a 'belated recognition' (Shearing 1997) of what has been a well-established practice by private actors to govern through others. In the following passage Shearing and Stenning suggest that corporations understand indirect governance to be more instrumental in achieving their objectives than direct governance through reactive 'bandit-catching' measures:

> The public police, with their apprehension orientation, have tended to direct their surveillance at potential troublemakers, that is, to what Spitzer (Spitzer 1975) has called 'problem populations'. While this is undoubtedly an activity that private security persons also engage in, for instance in the detection of shoplifting (Jeffries 1977), this focus on breaches of the law is only one aspect, and probably the least important one, of private security surveillance. Private security's emphasis on prevention directs its surveillance not so much to breaches of the law (or of organizational rules) as to *opportunities* for such breaches. As a consequence, the objects of private security surveillance tend to be not just potential troublemakers but also those who are in a position to create such opportunities for breaches. Thus, the target population is greatly enlarged. For example, within a business setting the focus of surveillance is not simply potential rule-breaches but anyone who might contribute to the creation of an opportunity for a breach of a rule. (Shearing and Stenning 1983b: 98)

The authors go on to illustrate indirect governance by corporations through the example of a 'snowflakes' policy that formed part of a security company's evening patrol routine:

> For each risk found, the patrolling security officer fills out and leaves a courteous form, called a 'snowflake', which gives the particular insecure condition found, such as personal valuable property left out, unlocked doors, and valuable portable calculators on desks. A duplicate of each snowflake is filed by floor and location, and habitual violators are interviewed. As a

> last resort, compliance is sought through the *violator's* department manager. (Luzon 1978: 41, cited in Shearing and Stenning 1983b: 99, their emphasis)

What differentiates public from private third-party policing is the nature of the authority of the enrolling actors. The public police carry the legal authority of criminal law in their crime control activities. In contrast, private third-party activities, such as those carried out by insurance companies, derive their authority from contract law (Roach Anleu *et al.* 2000: 79).

There are other bases of private authority as well. Authority can reside in markets themselves (see Cashore 2002; Cashore *et al.* 2004). This can be seen in many fields of governance beyond the security domain. For instance, in their analysis of the governance of intellectual property, Drahos and Braithwaite (2002) demonstrate very clearly how large corporations have used states, in particular the US, as vehicles for acting at a distance in their efforts to establish the conditions for a global knowledge economy that will deliver enormous profits to them. They show that the United States is enrolled precisely because it is such a powerful state in engagements with other states. Through enrolment US corporations gain a powerful ally in their attempts to promote their agendas globally.

The rise of private authorities is just as central to recent analyses of global governance as it is to local and national governance. Global authority is now exercised by actors such as business corporations, human rights and environmental non-governmental organizations, religious movements and mercenary armies (Hall and Biersteker 2002a: 4). Hall and Biersteker explain:

> While these new actors are not states, are non state-based, and do not rely exclusively on the actions or explicit support of states in the international arena, they often convey and/or appear to have been accorded some form of legitimate authority. That is, they perform the role of authorship over some important issue or domain. They claim to be, perform as, and are recognized as legitimate by some larger public (that often includes states themselves) as authors of policies, of practices, or rules, and of norms. They set agendas, they establish boundaries or limits for action, they certify, they offer salvation, they guarantee contracts, and they provide order and security. In short, they do many of the things traditionally, and exclusively, associated with the state. (Hall and Biersteker 2002a: 4)

It is thus never simply, as the Hobbesian perspective assumes, the concerns and objectives of one set of auspices, namely states that shape governance. While states seek to realize their programmes by mobilizing the knowledge, capacities and resources of others, other auspices are clearly acting in very similar ways to realize their agendas.

Nodal governance

In our discussion we have suggested that governing auspices are both objects of governance (by those who seek to enrol them) and actors that govern directly or through others. The language of 'nodal governance' endeavours to capture this complexity. This conception is meant to extend the insights on 'multilateralization' (Bayley and Shearing 2001) and 'pluralization' (Loader 2000; Johnston and Shearing 2003; Shearing and Wood 2003b; Burris *et al.* 2005) and locates itself squarely within the conceptions of power and governance expressed by Latour and Foucault. It focuses its analytical attention on the mentalities, institutions and practices of governing entities, or nodes, as well as the ways in which nodes may form governance relationships with others. Indeed, nodes may not come together to form 'networks' at all.

Rather than assuming that nodes are networked, a nodal framework regards this as an issue to be explored empirically. Nodes are sites of knowledge, capacity and resources that function as governance auspices or providers. These sites are often institutional (expressed in an organizational form), but can also be located within informal groupings. Consider Burris' description:

> [A] node must have some institutional form. It need not be a formally constituted or legally recognized entity, but it must have sufficient stability and structure to enable the mobilization of resources, mentalities and technologies over time. A street gang can be a node, as can a police station or even a particular shift at a firehouse. A node like this may be primarily part of an integrated network, like a department in a firm; it may be linked to other nodes in multiple networks without having a primary network affiliation, like a small lobbying firm; or it may be what we call a 'superstructural node,' which brings together representatives of different nodal organizations ...

> to concentrate the members' resources and technologies for a common purpose but without integrating the various networks – a trade association, for example. (Burris 2004: 341–2)

An important element of Burris' description is the idea that nodes may be linked to one another through networks and assemblages but they also may not be. They may have many links, but may also only have a few. In developing a nodal governance framework we are thus seeking to provide a conceptual architecture for analysing trends in governance that is equally comfortable with the idea that governance can be contested and uncoordinated as it is with the idea that it can be cooperative and coordinated.

Where cooperation does take place this may be state-orchestrated. There are, however, many other sources and forms of coordination. These centres may work with states but they can also coordinate nodes to resist and contest state governance. A broad 'mapping' that we undertook along with colleagues on policing and security nodes within Canada and internationally concluded with the following observation:

> The nodes that take part in nodal governance relate to one another in a wide variety of ways, which are characterized by differing degrees of official and tacit coordination. In some cases, network activities and outcomes are determined by a very small number of dominant nodes. The influence of such dominant nodes can be accepted and resisted by other nodes to varying degrees. In other networked arrangements, there is no conscious coordination of nodal activity – yet even in such instances of 'benign neglect,' the activity of individual nodes cannot help but impact upon the activity of the others, shaping the character of the overall network of governance that emerges. (Hermer *et al.* 2005: 39)

What we do know, particularly from Dupont's studies, is that established state nodes, particularly public police organizations, are increasingly cognizant of the multiplicity of auspices and providers operating in the field of security and are aware that they must actively 'jockey for position' if they wish to maintain a privileged place within this field. Based on interviews with senior police managers in Australia, Dupont found – drawing on the thinking of Bourdieu (1986) – that public police are both deploying and mobilizing various forms of 'capital' – economic, political, social and cultural – in order to maintain a certain degree of hegemony. While private security providers are accumulating different forms of capital – particularly of an economic nature – police managers are, to a certain degree, 'hyping up' what they claim to be those elements that continue to distinguish the public service; that is, its high levels of legitimacy and integrity as well as its high levels of specialist knowledge and capacity (cultural capital) (Dupont 2003, 2006a).

The nodal perspective attempts to account for the issue of 'mission creep' involving a blurring of functional specialization that in times past had served to distinguish between the missions and roles of different nodes, particularly with regard to their 'public' and 'private' character. Part of this functional blurring is a blurring of mentalities.

Previous literature on the governance of security, including our own, has tended to suggest that private nodes govern in ways that are distinct from public ones (Shearing and Stenning 1983a; Shearing 2001a). This has been attributed to the different sets of objectives and interests guiding public and private governments. In schematic terms, it has been argued that state authorities operate according to a 'punishment mentality' (Johnston and Shearing 2003), which denotes a backward-looking orientation focused on redeeming the past through the righting of wrongs. In contrast, corporations are said to operate through risk, denoting a more forward-looking orientation focused on shaping the future by manipulating present flows of events (Johnston and Shearing 2003). A nodal governance perspective, however, serves to confound this rather neat distinction between ways of thinking across 'public' and 'private' nodes':

> Not only is it very difficult to encounter what was once thought of as 'an essential state police function' that is not in many cases also performed by non-state policing agencies (with or without state approval and/or tolerance), but also the sharp demarcation often drawn between the preventive mentality of non-state policing and the past-oriented 'justice' orientation of state police is increasingly dissolving. The evidence to date indicates that innovations in ways of thinking about – and approaching – the problems of policing have frequently originated in the non-state sector, and have subsequently been diffused to the state police ... Conversely, though to a lesser extent, non-state policing agencies have in some instances begun to concern themselves with carrying out a past-oriented 'justice' function, mimicking, in so far as the law permits, the coercive capacity and symbolic

> authority of state police to the extent that they have discovered this to be a saleable service that the public demands. (Hermer *et al.* 2005: 49)

Braithwaite makes the more fundamental point that scholars of policing, and criminologists more generally, have largely focused their attention on mentalities of punishment because such scholars have paid most attention to the institutions of criminal justice at the expense of significant developments in other regulatory institutions and contexts (Braithwaite 2003a). He reminds us, as does Garland, that the concept of 'police' in pre-modern times referred to a broad domain of 'regulation, inspection and restraint' (Garland 1996: 464). 'Police regulations' covered a huge field of conduct, ranging, for example, from religion, to customs, to health, to food, to commerce to 'tranquillity and public order' (Pasquino 1991: 110). This 'science of police', in Pasquino's view, consisted of 'a great effort of formation of the social body, or more precisely an undertaking whose principal result will be something which we today call society or the social body, and which the eighteenth century called "the good order of a population"' (1991: 111). The institution of police at that time, according to Braithwaite, 'was rather privatized, subject to considerable local control, heavily oriented to self-regulation and infrequent (even if sometimes draconian) in its recourse to punishment' (Braithwaite 2003a: 9).

While Peel's new police were originally intended to be preventative, they gradually emerged as an institution of reactive punishment. This shift towards a punishment mentality was, according to Braithwaite, 'partly a Benthamite project of utilitarian rationalization and partly a political project of asserting central control over the disorderly populations of the industrializing metropolis where revolution was most feared' (2003a: 10). He asserts that this growing focus on regulating citizens on the streets did not mean, however, the end of regulation of other areas of life such as the regulation of business. For instance, from the middle of the nineteenth century onwards, the establishment in many nations of the 'public police' as a specialist regulatory institution was accompanied by the development of various other specialist regulatory institutions, including liquor licensing boards, mines inspectorates, health and sanitation boards and food inspectorates (2003a: 10). As Braithwaite explains:

> Unlike police forces, these regulatory institutions mostly started out rather punitive, making heavy use of criminal punishment as a regulatory tool, but mostly became much less punitive over the next 150 years ... The business regulatory agencies grew to be more significant law enforcers than the police because the corporatization of the world in the 20th century changed the world to a place where most of the most important things done for good or ill in the world were done by corporate rather than individual actors. (Braithwaite 2003a: 10–11)

Those who have studied mentalities of business regulation have found evidence of a variety of mentalities in operation at the same time. For example, some regulatory institutions continue to support the Hobbesian view that corporate actors are hedonistic and amoral calculators who will only comply with laws in the face of command-and-control measures designed to punish and deter. At the same time, however, other regulatory institutions operate on the assumption that 'virtuous' corporate actors do exist who are 'reasonable, of good faith, and motivated to heed advice' (Braithwaite 1989: 131). For Braithwaite, then, the field of business regulation informs us that non-punitive mentalities have come to significantly influence the punishment orientation of criminal justice institutions:

> The audit society ... actuarialism ... situational prevention, risk analysis, empowering victims, partnership, public/private alliance, co-production of security ... are actually best understood as products of the interplay between markets and the institutions created from the mid-19th century (and earlier) to regulate them. (Braithwaite 2003a: 12)

While corporate governance, and private governance more generally – through mechanisms such as private security – has emerged as a significant feature of nodal governance arrangements, the role played by these auspices and providers in the arena of plural governance constitutes a well-developed area of inquiry within the security literature (see, for example, Spitzer and Scull 1977; Shearing and Stenning 1987; Shearing 1992, 1996; Jones and Newburn 1998; Kempa *et al.* 1999; Johnston 2000a, 2006b; Mopas and Stenning 2001; Mandel 2002; Rigakos 2002; Avant 2005; Johnston and Shearing 2003; Singer 2003; Cooley 2005; Crawford *et al.* 2005; Crawford 2006; Jones and Newburn 2006). In this volume we focus our attention on two less well-researched features of nodal security governance – the nodal location of state policing and 'human security'.

While state policing has been and continues to be the most developed area of thinking within security governance, what remains relatively neglected is the shifting role of the police as they recognize and respond to plural governance. Another under-researched area in studies of security governance is the domain of human security, which has come to the forefront of much thinking over the past several decades. It is these two foci that are central to our discussions in this volume. This focus should not be read as implying that we do not regard other fields of inquiry within security governance – in particular, private governance – as unimportant, as our own previous work should make clear (see recently Shearing 2006). These are simply not areas of enquiry that we have chosen to focus our attention on in this volume.

As a corrective to earlier studies of security governance – particularly those focused on the phenomenon of private security – scholars are placing more attention on the ways in which mentalities mix and meld, particularly those based in punitive and risk-based orientations (Johnston 1997, 2000a, 2006b). This melding can be seen very clearly in public policing, which is something we explore in detail in Chapter 2. At the same time, the influence of punishment mentalities has not diminished. Punishment as a way of acting on subjects or targets of governance is being expressed and carried out in new ways by a plurality of governance nodes including but not limited to commercial military service providers, as we mention in Chapters 3 and 5.

There is more to nodal governance, however, than the melding of punitive and risk-based thinking within and across state and corporate nodes. One can also observe the emergence of new auspices and providers of security from Drucker's 'third sector' (Drucker 1994). These auspices include non-governmental organizations working in the spheres of development and human rights. As evidenced in the 'human security' movement, the focus of discussion in Chapter 3, such organizations express particular security mentalities and bring to bear new forms of knowledge, capacity and resources to the field of security provision. In so doing, they are actively shaping this field through attempts to shift beyond a state-centric conception of security, and the Hobbesian mentality that traditionally underlies it, towards a human-centric notion that articulates a wide range of threats to the material existence of human beings and to their personal, social and economic development. In 'securitizing' problems previously conceptualized within other domains of governance, new auspices and providers of human security governance are forming various 'strategic alliances' with traditional security institutions and with the corporate sector (Duffield 2001).

A nodal governance analysis seeks to understand the various mentalities and strategies of nodes and nodal arrangements that operate at the same time. The findings of such explanatory work have normative implications. We explore these in Chapters 4 and 5. Chapter 4 examines the 'governance disparity' between 'weak' and 'strong' actors in the field of security. In jockeying for position in the security field and in enrolling others in security networks, not all nodes are equally capable and effective. It is this difference that we highlight and we offer some tentative ideas as to ways in which the capacity of weak nodes can be enabled and developed in order that they can more actively direct the provision of security goods.

Chapter 5 explores the challenge of governing nodal security provision. Traditionally, the domain of regulation or the 'governance of governance' has been conceptualized from a state-centred point of view. That is, attention has been focused on governing state providers, particularly public police, to ensure that they conform with normative standards of conduct in areas such as the lawful use of force and due process. In this way, the governance question has been framed largely as a question of 'police accountability' and of the capacity to provide retrospective accounts of police actions to legal and political authorities if breaches of professional police conduct are identified. In our examination of the 'governance of governance' we look at ways in which the governance of public policing has itself been 'nodalized' through the introduction of new mentalities, particularly those originating in the private sector. We go on to argue, however, that the potential for the *nodal* governance of governance has not been fully realized due to the strong foothold that state-centred explanations of security provision and regulation have retained (Lewis and Wood 2006).

Conclusion

The shift to a nodal governance perspective that we advance in this book is not one that assumes a decline in state authority and power. Indeed, one could argue that state power is now even more diffuse and pervasive through the ways in which it governs through the knowledge, capacity and resources of others. The conceptual shift we advocate is simply one that recognizes the diversity of entities, what we term 'nodes', that function as auspices or providers of security goods or both. For us '[t]he state is no longer the sole, or in some instances even the principal, source of authority, in either the domestic arena or in the international system' (Hall and Biersteker 2002a: 5). The significance of this for our understanding of governance is that it is not simply some of the rowing of governance that has shifted from state to non-state entities; such shifting has applied as much to the steering of governance. The presence of private auspices has not simply 'reinvented government'; it has 'reinvented governance'. In doing so it has greatly expanded the range and scope of non-state interests in the delivery of security goods.

From a nodal governance perspective, one finds a diffusion of mentalities – ways of thinking – across nodes. There may indeed be ways of thinking that are 'natural', or at least dominant, within particular nodes. This is argued to be the case with respect to the 'punishment mentality' of criminal justice and risk-oriented thinking among corporate auspices (Johnston and Shearing 2003). However, it is increasingly the case that mentalities from some spheres are influencing those of other spheres, serving to produce new forms of alignment between different ways of thinking (Johnston 1997, 2000a). In Chapter 2 we explore, through an analysis of 'waves' in public policing, ways in which punitive and risk-based thinking are aligned in attempts to govern through 'community'. Chapter 3, on human security, then examines efforts to introduce new mentalities into the mix.

Chapter 2

Community security and local governance: waves in public policing

Introduction

In the previous chapter we described nodal governance as a conceptual frame for understanding the contemporary field of security and changes that have been taking place within it. Consistent with analyses provided by Johnston (Johnston 1997, 2000a; Johnston and Shearing 2003) we argue that a nodal governance perspective allows one to observe ways in which the thinking of one node or set of nodes shapes and penetrates the thinking of other nodes. This is particularly important as any study of changes in governance strategies and practices must involve an analysis of the mentality shifts that underlie them. Developments in public policing over the past quarter of a century provide an illustration of such mentality shifts. In this chapter we seek to advance the idea that mentality shifts occur not simply as leaps in thinking as suggested by the notion of 'paradigm' shifts. Instead we conceive of change in terms of 'waves' (Wood 2000; see Toffler's conception, 1980).

A 'waves-based' view recognizes multiple, and often overlapping, mentalities that have emerged across times and places. Some waves emerge as an extension of ideas introduced through other waves, while other waves emerge as a consequence of resistance to established ways of thinking. Through an analysis of waves in the realm of public policing we aim to enhance our understanding of how public policing has been and is being re-imagined, as well as the conditions that have allowed for these imaginings to come into and out of view. As sociologists have long pointed out, the public police bring unique capacities in the governance of security, in particular the exercise of non-negotiable force through law enforcement (Klockars 1985; Bittner 1990). The innovations we examine in this chapter are understood as responses to critiques of and/or failures in the ways in which these capacities have been thought about, managed and exercised within established liberal democracies. One element of these critiques has focused on nodal relationships – specifically the linking of police knowledge and capacities to those of non-police nodes. We view each wave as attempting to align new ideas pertaining to 'community' and community capacity with views about the police and their strategic and symbolic place within the field of local security delivery.

In examining waves in policing, our focus is on the ideas that have become 'thinkable' or 'imaginable' – particularly those pertaining to the linking of police and non-police capacities – and possibilities for generating new waves of thought. Before we turn to our analysis, we pause briefly to comment on the police identity and role as traditionally conceived within a state-centred framework.

The place of the police

As we have noted, the public police were established as a quintessential Hobbesian institution. Their capacity to exercise non-negotiable force has been and continues to be essential to their identity and role. While the police constable may choose not to exercise these capacities in the majority of situations, the threat of coercion constitutes a key component of their legal and symbolic authority (Shearing and Leon 1977: 342; Bayley and Shearing 1996: 21; Loader 1997). This recourse to coercion is not a unique feature of the police role or the military as governments can and do authorize others to use non-negotiable force (see subsequent chapters). Weber puts this argument about the centrality of force to state institutions this way:

> … force is certainly not the normal or only means of the state – nobody says that – but force is a means specific to the state … the state is a relation of men dominating men [and generally – one should add – men dominating women], a relation supported by means of legitimate (i.e. considered to be legitimate) violence. (cited in Held 1983: 35)

This conception of coercion is supported by a 'punishment mentality' that has provided a long-held set of instrumental and symbolic (moral) justifications for its use as a governance technique (Johnston and Shearing 2003). This mentality is based on a Hobbesian conception of human agency that understands human beings as possessing selfish desires or passions that can only be effectively regulated by a strong sovereign power standing over and above people. Because we are also rational, choice-making actors, we see it as in our interests to surrender certain freedoms, through compliance with universal laws, in order to avert a 'war of all against all'.

Beccaria (1764), in extending this Hobbesian view, argued that punishment, when deployed appropriately and proportionately, promotes the reduction of crime through the mechanisms of specific and general deterrence. This 'consequentialist' (instrumental) view of punishment (Johnston and Shearing 2003: 40) is aligned with a symbolic, 'non-consequentialist' view that punishment is also justified as an end in itself – 'the justification of punishment is considered to be its inherent "rightness" vis-à-vis the (past) offence, rather than any future (deterrent or other) value it might possess' (Johnston and Shearing 2003: 40). This conception of the role of punishment underlies established conceptions of the place of police within the governance of security.

In English-speaking liberal democracies, Sir Robert Peel's imaginings of public policing continue to shape innovations in the local governance of security. In the spirit of Peel's dream to professionalize and centralize activities previously left to communities, modern police institutions emerged to promote the 'social' and the sovereign nation as the imagined basis for collective life. Associated with this understanding of collective life was a conception of 'security' as a 'public good' (Loader and Walker 2001). While 'just a citizen like you' (De Lint 1997: 250), Peel's constable represented an impartial 'public' and not a partisan police. Similarly, unlike citizens, police were conceived of as professional (experts) in the promotion of security. In order to support this non-partisan and specialist understanding, policing would be paid for, directed by, and accountable to the state.

Peel's new professional police was established as a hierarchical organization that managed itself through systems of command and control based on the rule of law (Uglow 1988: 21). This institutional arrangement – characteristic of all state bureaucracies – would not only house the vast amounts of professional expertise, but would be characterized by 'military personalities and military structures' (Task Force on Policing in Ontario 1974) seen as essential to the legitimacy of an institution of 'domestic specialists in the exercise of legitimate force' (Reiner 1992: 762). Indeed these structures and identities were considered to be essential if police officers were to function as 'impersonal professionals' (Crawford 1997: 20).

The purpose of Peel's police was to *prevent* breaches of security by establishing an 'unremitting watch' or 'panoptic gaze' (Shearing 1996: 74; Foucault 1977) through 'visible uniform patrolling of the streets' (Crawford 1997: 19). Within this system of surveillance, citizens would be 'deterred' from breaching security because they would know that their chances of getting caught by the police, and being punished, would be high (Shearing 1996: 74). This preventative conception, however, became difficult to realize in practice for two key reasons. First, the 'institutions of privacy' (see Stinchcombe 1963; Shearing 1996: 74–5) associated with liberal states restricted the police to 'public' spaces and they, accordingly, did not enjoy unrestricted, routine access to private spaces. To cope with these restrictions, police developed strategies over time to govern through others, especially ordinary citizens, by enrolling them as 'watchers', collectors and reporters of private information (Shearing 1996: 75).

The second major impediment to the Peelian dream of prevention came in the form of pressure on the police to demonstrate 'visible outcomes' (Crawford 1997: 20) of their activities. With advances in the growth in statistical techniques, indicators such as arrest and crime rates became increasingly valued as measures of police performance. To this day, police organizations are required to adhere to a range of 'reactive' performance indicators, such as arrests and clearance rates, as key markers of their effectiveness. While this is changing in line with a new 'customer service' orientation (see Chapter 5), there remains considerable pressure to measure security according to indicators of crime and criminals.

Up until the mid-twentieth century, the various shifts that took place in western industrialized democracies that served to decentre the state did little to alter this conception of the police role. In fact, much of what occurred – through further bureaucratization and militarization – served to embed more firmly the established 'governing through crime' conception through which 'crime and punishment become the occasions and the institutional contexts in which we undertake to guide the conduct of others (or even ourselves)' (Simon 1997: 174). In what follows we examine, using Ontario, Canada as an exemplar, the ways in which new ideas challenging the governing-through-crime orientation were taken up by public police organizations in 'waves' of policing (this discussion draws extensively on Wood 2000).

Waves in public policing

During the last quarter of the twentieth century the place of the police, in both symbolic and instrumental terms, came under increased scrutiny by academics, politicians and policy makers. At an instrumental level, critical assessments had been mounting regarding the limits of a punitive, reactive policing style in the production of local security. At the same time, police organizations in countries like Canada were beginning to witness significant shifts in the social and cultural make-up of the populations they served. With an increasingly diverse public, the challenge for police organizations was to rethink how they could embrace diversity while at the same time retaining the image of the police as iconic representatives of the public interest.

Policing as community-based

Bayley described the first wave of 'community policing' as 'old wine in new bottles' (Bayley 1988: 225–6). During this wave police reformers sought to re-imagine policing as an essentially 'community-based' activity – one that fostered symbolic and strategic links with everyday citizens. They sought to reconfigure the established police identity as detached professionals, and to rethink the Peelian image of the 'police as the public and the public as the police' in the face of increasing social diversity and cultural heterogeneity. In Crawford's words, police sought to 'reconfigure the Peelian legacy in order, somehow, to reconstitute physical and psychological relations between the police and public, to build trust, and to encourage greater public assistance in policing' (Crawford 1997: 45).

In this re-invocation of the Peelian legacy there were shifts in the language and imagery deployed in police management texts, policy statements and political announcements. In Ontario, police organizations themselves argued that police needed to become more *sensitive* to the changing interests and requirements of the population while at the same time maintaining their 'traditional responsibilities of protection and law enforcement' (Ontario Provincial Police 1989a: 2). Within this conception, to be legitimate agents of law enforcement required police to 'be in harmony' with those they served (Ontario Provincial Police 1989b).

Police reinforced their identity as icons of sovereign authority while becoming more community-*oriented* and community-*based*. To realize (and present) this changing face of policing, this wave saw the introduction of 'community policing specialists', such as 'community police officers' in places like schools (1989b) and storefront police offices, who provided traditional policing services with a public relations twist. Police were now to be 'more engaged in the community through programs designed to inform and direct community involvement in law enforcement' (Ontario Provincial Police 1995a: 3).

Community involvement in professionalized policing was regarded as an important value on its own (Tilley 2003: 326), regardless of the knowledge and capacities they brought to bear on issues of local safety. Consistent with the Peelian dream, the police sought assistance and cooperation, but themselves remained expert diagnosticians of community problems. This privileging of police capacity reinforced the broader view of state-centred governance rooted in their specialist knowledge and capacities as well as their paternalistic relationship to citizens and communities (O'Malley and Palmer 1996).

In this wave the language of 'community' is mobilized as a means of imagining collective interests in the face of growth in the cultural and demographic plurality of the citizenry. 'Community' denoted 'mini-socials'. In Ontario, a means used to align an emphasis on social diversity with a commitment to impartiality and universality was to emphasize the principle of *equity* through the establishment of programmes, both inside and outside of the police organization, to give this principle concrete effect. One approach to this was to build in an Employment Equity Plans Regulation into Ontario's Police Services Act. This required all provincial policing services to develop an Employment Equity Plan for eliminating discrimination and for promoting equality in hiring practices (Ontario Provincial Police 1991: 15; Ministry of Solicitor General and Correctional Services 1993: 3). Another example was the establishment of the Race Relations Policy for Ontario Police Services – 'policing in our province [should be] equally responsive to all Ontarians without regard to race or colour' (Ministry of Solicitor General and Correctional Services 1993: 1).

During this wave, the introduction of new language and imagery ('equity', 'harmony' and 'sensitivity') constituted an important alignment strategy. Professional policing was retained as the central operational paradigm, but with a new gloss, aptly described by Deukmedjian as 'community relations policing' (Deukmedjian 2002: 101).

This wave also brought with it an important discursive shift when it introduced the new language and imagery of policing as a *service*. What was involved here was a symbolic shift in the conception of police, from an organization centred on the application of force to one focused on service. In Ontario, as in many other jurisdictions, the title of the legislation governing all provincial police organizations was changed from the Police Act of Ontario to the Police Services Act of Ontario (Police Services Act, R.S.O. 1990, c. P-15). Individual organizations embraced this new image of themselves as a *service* rather than a *force* (Miller 1996). This re-imagining was part of the same alignment project as the shift to 'equity' and 'sensitivity'. Policing was still conceived as a public good but there was a shift in the way the public were conceived. They were now imagined as 'clients' with diverse needs and expectations. To reconceive policing as a service was to acknowledge a diversity of needs and expectations without at the same time favouring particular needs and expectations over others. Policing was to be fair and equitable, rather than 'tailor-made', as was implied within the subsequent market-based conception (see below).

In summarizing this wave, we draw from Manning:

> Rather than being a new strategy that replaces the crime control professionalism strategy that produced social distance, [community policing] is a contrapuntal theme: harmony for the old melody. It now seeks control of the public by a reduction in social distance, a merging of communal and police interests, and a service and crime control isomorphism. (Manning 1988: 28).

Whether or not one sees this wave as 'old wine in new bottles', it served to open up new conceptual spaces within which to re-imagine the instrumental and symbolic relationships between the police and other actors.

Policing as solving problems

Part of this first wave of community policing was symbolic work carried out to relegitimize the institution of the public police as the central guardians of public security. While the instrumental features of professional policing remained intact during that wave, the limitations of the 'governing through crime' approach were being increasingly recognized by scholars and practitioners. Supported by early research findings, such as those produced by the Kansas City study of patrol effectiveness (Kelling 1974), an increasing number of academics and practitioners began to conclude rather cynically that crime often *does* pay (Bayley and Shearing 1996: 588). This recognition was one of the precursors of a second wave of change within state policing.

A key driver of this next wave was the argument that the mobilization of police capacity contributes surprisingly little to the production of community security. Goldstein, the central intellectual figure in building this critique, argued that the ineffectiveness of the police could be attributed to a conflation of the 'means' and 'ends' of policing, which he characterized as the 'means over ends syndrome' (Goldstein 1979). While the police officer may not have been spending the bulk of their time catching bandits, the police organization was structured and managed as if this was the officer's main objective. Goldstein argued in particular that while the police possessed the authority to enforce the law, this did not mean that law enforcement was, or should be, their primary objective. Rather than seeing themselves as 'law enforcers', the police should re-imagine themselves as 'problem-solvers'. The job of the police, he argued, 'requires that they deal with a wide range of behavioural and social problems that arise in a community – that the end product of policing consists of dealing with these *problems*' ([1979] 1991: 483, italics in original). Goldstein did not expect the police to *solve* or *eliminate* the 'residual problems of society'. Rather, he argued, 'it is more realistic to aim at reducing their volume, preventing repetition, alleviating suffering, and minimizing the other adverse effects they produce' (Goldstein [1979] 1991: 483). It is during this wave that the language of 'problem-oriented policing' (Goldstein 1979, 1990) was established. As Bradley explains:

> ... problem-oriented policing would see most police officers operating in highly autonomous working environments. They would see their primary core business as generating data about incidents and cases and looking for possible relationships between such cases. They would seek to identify the basic underlying problems indicated by such relationships, and more accurate and comprehensive way[s] of describing them, analyse their causes, and then set about tackling them. Problem-solving would in the first instance always look for ways in which non-police agencies and efforts might wholly or in part provide solutions or amelioration of the outcomes, and, feeding on this, the cycle would start again. *If there is one primary distinguishing characteristic of problem-oriented policing it is its focus on broadly defined social outcomes of policing activity, in contrast to a narrow concern with legally-defined process and criminal law enforcement as an end in itself (although this is not to say that it in any way abandons the notion of due process).* (Bradley 1994: 2; emphasis in original).

Goldstein's critique addressed the 'governing through crime' mentality. Not only was the language and imagery of crime and criminal justice problematic in 'making up' the problem of security, the highly reactionary nature of traditional policing meant that the wide-ranging social problems the police came into contact with were not addressed adequately. In response to this, Goldstein was calling on the police to inject more future-oriented thinking, and in particular to place more of an emphasis on the risks posed by various social problems. In order to achieve this, the police would need to see law enforcement as simply one among a variety of means they could deploy in managing these risks.

Since the central objective of this problem-oriented policing approach was the management of social problems, the means essential to the police (coercive force and law enforcement) were regarded as instrumentally limited as governance technologies. During this wave police organizations began to take up Goldstein's language and re-imagine police officers as multi-capacity agents (De Lint 1997), who would 'anticipate future calls by identifying local crime and disorder problems' (Ministry of Solicitor General and Correctional Services 1991: 7).

A crucial implication of this imagining was that it required police to link up with others who had the knowledge about the 'underlying causes of problems' (1991: 7). Within this conception it is no longer adequate to think about community policing as an organizational 'add-on', or as a 'special program to be added to the way policing is conducted' (1991: 5). In place of the specialist 'community police officer' is now the image of the 'generalist community policing officer' (Wood 2000) who, as De Lint explains, is 'more than simply a law enforcer'. This re-imagining of police officers conceives of them as 'exemplif[ying] social norms and [as] an iconic representation of good citizenship; more than just an order maintainer, she uses networks and agencies to proactively re-create secure environments; more than just a social worker she diagnoses social problems and attacks root causes' (De Lint 1997: 248).

To re-imagine the police officer in this way generated new alignment challenges. The very structure, management and culture of the police organization would need to be aligned with this broader constabulary function (Ministry of Solicitor General and Correctional Services 1991). Herein rested a tension; the police were to remain agents of law enforcement and legitimate bearers of coercive force. A paramilitary organization and managerial structure were still regarded as required to govern the exercise of coercion. At the same time, however, if the police were also to be understood as a problem-solving agency, a flatter, more horizontal and decentralized organization was required in order to enhance the 'responsibility and autonomy' of officers (1991: 9).

This alignment challenge was articulated in police reform documents, such as the 1974 report by the Task Force on Policing in Ontario. Rather prophetically, this report suggested that innovations in one aspect of policing, such as the introduction of a new problem-solving ethos among officers, would not take effect if other aspects of policing were left untouched, such as its paramilitary organization. Perhaps most significantly, the report warned that policing, if not reorganized, could not be sustained economically (Task Force on Policing in Ontario 1974). And indeed, as became clear during the late 1980s and early 1990s, economic imperatives prompted yet another nodal re-imagining of policing for these reasons.

The influence of neo-liberalism

In Ontario, as elsewhere, this continuing quest to re-imagine policing was driven significantly by fiscal reasons – what was dubbed at the time as a fiscal crisis (Wood 2000). This was particularly true in the United States, where it became increasingly unthinkable to do 'more of the same, in the form of more hiring of officers to conduct policing in the traditional style. Both police and their political masters [were] looking for ways to get more out of less' (Skogan and Hartnett 1997: 10). For the Ontario Provincial Police, policing became viewed as unnecessarily 'hampered' – a conception that resonates with the thinking of the 9/11 Commission cited earlier – by an institutional framework 'that was more consistent with earlier generations of police service delivery models' (Ontario Provincial Police 1995b: 1).

With the widespread agreement that states faced a crisis around the globe in the early 1990s, a neo-liberal mentality took hold that argued that sensibilities traditionally associated with the corporate sector should be taken up by government bureaucrats and police managers and melded with discourses of problem-oriented policing (see Chapter 5). Police organizations were encouraged to re-imagine themselves in business-like terms (O'Malley 1997), seeing themselves as delivering 'products' in the most cost-effective manner as possible. Because problem-oriented policing was a strategic vision that called for a new image of police organization – leaner, flatter, team-oriented – it was a rather straightforward conceptual move to align the image of problem-oriented policing with that of a 're-engineered' organization. Neo-liberal state reforms throughout the world (Rose and Miller 1992; O'Malley and Palmer 1996; O'Malley 1997) involved 'streamlining' processes that did not 'add value' or that produced inefficiencies.

With the incorporation of mentalities from the corporate sector, problem-oriented policing (POP) was refigured as an approach that promoted the more effective use of police resources. In the short term, problem-solving processes would encourage police officers and those with whom they were networked to explore cheaper alternatives to criminal justice processing in ways that resonated with a POP orientation. In the longer term, it was argued, resources would be saved from not having to deal with problems that escalated and became a part of the 'revolving door of criminal justice'. At the same time, POP was conceived as entirely consistent with the symbolic shift in the previous wave towards an image of police officers as sensitive and equitable service providers. Indeed, problem-oriented policing was understood as deepening the commitment to building relationships with others, since it was now essential to mobilize and link with a wide range of non-police capacities. All of this contributed to creating a conception of state-anchored nodal policing as the best way of approaching the governance of security.

During this wave police began gradually to draw back from the language of 'professional' or 'traditional' policing and replace it with language such as 'contemporary policing'. As part of this shift, the means essential to the police – non-negotiable force – came to be seen as only one among an array of means in solving problems (Ontario Provincial Police 1996, 1997, 2004: 1–2). Now members of the public were no longer simply imagined as cooperative providers of assistance, as in the 9-1-1 conception. The earlier paternalistic image of police–community relations was displaced by a new image of citizens and communities as 'active' and 'responsible' participants in security provision. Police organizations came to speak of communities as 'owners' of security problems (Ontario Provincial Police 1997). Within the Ontario Provincial Police community policing was spoken of as 'customized service delivery' (Wood 2000).

This shift to a conception of local security governance as a networked activity continued to be centred on specialist or professional forms of knowledge. One finds this in the language of new programmes for doing crime prevention through 'multi-agency' partnerships (Hughes 1996; Gilling 1997; Crawford 1998) and in the use of the phrase 'whole of government' to described networked governance. Within this shift the language of 'crime prevention' is significant. It does not call for a complete departure from the 'governing through crime' approach, insofar as it continues to understand the problem of security according to the language and categories of criminal justice. Security is still broadly conceived within the language of crime prevention as 'governing through crime' but with a risk-oriented twist.

During this wave there were, and continue to be, different approaches to the prevention of crime, each of which have had different implications for the place and use of coercive capacity. Each of these approaches has different implications for the way this wave of nodal policing was conceived. In the 'how to' manuals of community policing within the Ontario Provincial Police (1997) and elsewhere, we find reference to two ways of thinking about prevention: situational crime prevention and crime prevention through social development. Within the social development strand, two ways of thinking – control theory and strain theory – have well-established sociological roots. Control theory (Hirshi 1969) directs police to participate in nodal networks that enhance the internalization of external social norms. Within this conception schools and families are identified as important nodes that police should cultivate and build into policing (Einstadter and Henry 1995; Crawford 1998). Strain theory (Merton 1968), on the other hand, encourages police to participate in networks that operate to enhance opportunities that will lead people away from wrongdoing into more legitimate activities (Crawford 1998; Rosenbaum *et al.* 1998). While often perceived within sociology as opposing theories because of their differing conceptions of human agency, they both have been drawn upon in re-imagining policing within this wave.

While there are obvious differences between the situational and social developmental strands of crime prevention in regards to their theoretical lineages, there are also important points of convergence. Both strands of thinking, for example, emphasize the mobilization of professional knowledge and capacities through 'multi-agency' networks that locate the police as one professional node among others in ways that validate their specialized and professional knowledge base. Similarly, both strands of crime prevention agree on the need to focus resources on the management of risks in ways that resonate with the police focus on crime. While both strands agree that the police are not able to prevent crime alone, they both agree that police have a critical role to play in a 'joined-up' approach to governance.

Each of these approaches also encourages police to move beyond a 'whole of government' approach to facilitate and participate in 'whole of governance' or 'whole of society' networks that include non-state nodes. While police have not taken easily to this view they have nonetheless been moved in this direction as part of the wave of problem-oriented policing. This movement is seen most clearly perhaps in the police focus on building 'partnerships' across the state/non-state divide. There are many examples of this – crime stoppers, neighbourhood watch, business improvement arrangements, community policing committees, and the like. As this partnership conception has taken hold within the context of POP the police officer is being re-imagined yet again – this time as a partnership facilitator. Here the police are conceived of as catalysts and agenda setters in a field of networked policing. The notion of the police constable as a specialist in the use of force remains, but is now conceived as a resource that police have in addition to their role as a network coordinator.

The acceptance by police of new ways of thinking about the prevention of crime has served to open up space for re-imagining policing more generally. An important arena in which this space has been opened up is 'restorative justice', and it is to these developments that we now turn.

Policing as restorative justice

Restorative justice, specifically 'restorative policing' (Sherman 1999; Strang *et al.* 1999), picks up instrumental and symbolic strands from problem-oriented policing and crime prevention through social development. Like the social development strand it is concerned with 'root causes' and it understands coercive force as a necessary but nonetheless a 'last resort' in shaping the conduct of individuals.

For Braithwaite, 'Few sets of institutional arrangements created in the West since the industrial revolution have been as large a failure as the criminal justice system. In theory it administers just, proportionate corrections that deter. In practice, it fails to correct or deter, just as often making things worse as better' (Braithwaite 2000b: 319). Restorative justice (RJ) is described by Braithwaite as 'deliberative justice; it is about people deliberating over the consequences of a crime, how to deal with them and prevent their recurrence' (2000b: 324). It is during the wave of policing shaped through restorative justice that a fundamentally different conception of human agency is articulated.

RJ is based on a critique of state-based justice as alienating and frustrating for victims because the very structure of the process is based on a conception of the state as victim, rather than the individuals who have suffered a particular harm. According to RJ advocates, traditional criminal justice practices, drawing from Christie (1978), are designed to 'steal' the conflicts away from those affected by them (Braithwaite 2000b). It does so because 'the legal concept of guilt which guides the justice process is highly technical, abstracted from experience. This makes it easier for offenders to avoid accepting personal responsibility for their behaviour' (Zehr 1990: 72).

Advocates of RJ criticize traditional criminal justice on both instrumental and moral grounds. In line with the social development perspective, they argue that the conception of agency underlying the punishment mentality is limiting. They are critical of the idea that individuals are free moral agents. They are also critical of the non-consequentialist view that offenders must get their 'just deserts' (Zehr 1990: 70, 74) as if 'there is a metaphysical balance in the universe that has been upset and that must be corrected' (1990: 74). As a 'third model' of justice (Braithwaite 2000b), RJ promotes a future-oriented logic rooted in a different consequentialist argument (Braithwaite and Pettit 1990). According to Braithwaite (1989), crime prevention can be achieved through processes designed to repair the harm that was caused by the offender. This can be achieved not through punishment doled out by an indifferent party, but through a process of shaming the offender in a reintegrative rather than stigmatizing manner. Braithwaite's theory of reintegrative shaming (1989) is based on research which revealed that the societies with the lowest crime rate are those that are the most effective in shaming criminal conduct.

For Braithwaite, restorative justice calls for a de-centring of the police role, while ensuring that 'communities' (including friends, victims and other key reference groups for victims and offenders) play a central role. Nevertheless, RJ tends to work with the grain of established criminal justice processes (Shearing *et al.* 2006) as it promotes a process for diverting cases away from traditional sentencing channels to processes like 'family group conferences'. Furthermore, as Braithwaite argues, the coercive capacity of the police and that of criminal justice more generally remains as a governance alternative when there is a lack of compliance with the restorative approach. As he puts it, 'my hypothesis is that restorative justice works best with a spectre of punishment in the background, threatening in the background but never threatened in the foreground' (Braithwaite 2002: 35). Restorative justice is an emblematic example of what Braithwaite refers to as 'responsive regulation' (Ayres and Braithwaite 1992), where practices of punishment are applied only in response to the failures or inadequacies of informal or non-state governance processes (Braithwaite and Pettit 1990; Braithwaite 2002; Parker and Braithwaite 2003).

While working within the established paradigm of criminal justice, RJ rests firmly on the imagined space of 'community' based on a 'communitarian' vision. Communitarianism emphasizes universal agreement around particular normative standards, and in particular the 'clearly majoritarian morality' buttressing criminal law (Braithwaite 1989: 14). At the same time it calls on families, friends and neighbourhoods to take more responsibility in actions of socialization and reintegration:

> For a society to be communitarian, its heavily enmeshed fabric of interdependencies ... must have a special kind of symbolic significance to the populace. Interdependencies must be attachments which invoke personal obligation to others within a community of concern. They are not perceived as isolated exchange relationships of convenience but as matters of profound group obligation. Thus, a communitarian society combines a dense network of individual interdependencies with strong cultural commitments to mutuality of obligation. (Braithwaite 1989: 85)

In line with this communitarian thinking, RJ promotes the idea that the individuals most affected by a criminal act should be those most actively involved in addressing it. This should be accomplished through a deliberative process where families, friends and other individuals that constitute the 'community of care' of both the offender and the victim discuss the consequences of the crime and ideas for minimizing the risk that it will happen again (Braithwaite 1989). In this way, RJ aligns new instrumental and symbolic dimensions in that it reorients the police and other institutions of criminal justice towards risk management while at the same time articulating a new communitarian vision of collective life and collective responsibilities (Shearing 2001a).

The local knowledge and capacities of 'communities of care' supplement, rather than overturn, those of criminal justice professionals. As such, restorative justice programmes have often engendered a broadening of the police role. The typical, although not exclusive, family group conference structure includes a police officer in the role of facilitator (Moore and O'Connell 1994). Although emblematic of the 'spectre' of state punitive capacity, the police officer facilitates a deliberative forum designed to privilege local knowledge and capacity in addressing the harms caused by crimes.

In restorative justice we also see an integration of sensibilities from areas outside of criminal justice, particularly in the governance of corporations. Braithwaite's earlier work found that deliberative, 'conversational regulation' (Braithwaite 2003) was more effective than 'command and control' approaches to compliance with normative standards within corporate settings (Braithwaite 1985; Makkai and Braithwaite 1993, 1994). At the same time, Braithwaite traces restorative thinking back to ancient Arab, Greek and Roman civilizations (Braithwaite 2002). He goes so far as to argue: 'Restorative justice has been the dominant model of criminal justice through most of human history for perhaps all the world's peoples' (2002: 5).

The restorative justice movement is not the only wave in which we find a blending of risk-oriented thinking with a communitarian vision of collective life. Key features of 'broken windows' policing also embrace an understanding of risk management within a communitarian frame. Here, however, we find distinct imaginings of human agency and risk management that have different implications with respect to the forms of knowledge and capacities required to govern local security.

Policing as fixing broken windows

'Broken windows' has emerged as a strong policing wave because it resonates with strands of problem-oriented policing, risk-based thinking, as well as the broader communitarian currents to which restorative justice appeals. What is perhaps most significant about this wave is that it seeks to bring the police, and coercive capacity, *back in*, through the deployment of a conception of human agency wrapped within a risk discourse.

In line with problem-oriented policing, Wilson and Kelling (1982) argue that the police mission is to manage a wide range of social problems including, but not limited to, crime. These problems of 'disorder' include abandoned cars, graffiti, public drunkenness, street prostitution, youth gangs, public urination, and unlicensed peddling. Consistent with problem-oriented thinking, they argue that 'little things', if left unaddressed, may spiral into bigger things and become crimes. They argue that 'disorder and crime are usually inextricably linked, in a kind of developmental sequence' (Wilson and Kelling 1982: 31). That being said, the chain of causality – the relationship between disorder and crime – is different from that suggested by social development theorists.

The conception of human agency underlying 'broken windows' resonates more fully with the Hobbesian view underlying classical deterrence theory. Individuals are hedonistic, rational choice-making offenders who take advantage of opportunities to satisfy their selfish desires. Where this conception differs from the traditional Hobbesian view is in the weight it accords to communities in asserting normative standards of behaviour. For this conception it is not sufficient simply to ensure the presence of a sovereign authority, standing over and above society, to enforce social norms. A 'clearly majoritarian morality' must be reinforced not only through the criminal law but through the day-to-day practices of citizens within their neighbourhoods. For Wilson and Kelling, informal social control is just as, or more, important than formal practices of punishment. In nodal terms it is not simply the police node that is important.

From this perspective an important objective of informal social control is the maintenance of 'order', represented by the absence of physical and social markers of 'incivility' – rubbish strewn across streets is just as problematic as panhandling alcoholics. Both of these disorderly conditions signal that the (geographical) 'community' does not care enough to insist upon core normative standards of behaviour. The problem with not caring, according to Wilson and Kelling, is that disorderly conditions invite would-be criminals to engage in disorderly conduct and in some cases to commit more serious crimes. At the same time, community members become fearful of the very criminogenic conditions represented by the presence of physical and social disorder. Acting out of fear, citizens retract from social life and tend to stay in their homes rather than enter public spaces. The result of this is a community that 'spirals' into decline (Skogan 1990), inviting further 'criminal invasion' (Wilson and Kelling 1982: 32). '[I]f a window in a building is broken *and is left unrepaired*, all the rest of the windows will soon be broken … one unrepaired broken window is a signal that no one cares, and so breaking more windows costs nothing' (1982: 31, italics in original).

The implication of 'broken windows' for the police is that their coercive capacities are less needed in communities that already exercise strong informal social control. What is most important is for police to 'reinforce the informal control mechanisms of the community itself' (Wilson and Kelling 1982: 34). In this role, an important objective is to get to know both the 'regulars' and the 'strangers' in neighbourhoods, as well as disorderly people who live there but 'know their place' (1982: 30). Wilson and Kelling echo Goldstein in relation to the inadequacies of organizing police resources around criminal law enforcement (Goldstein 1979). They argue that the balance has been tipped too far towards the idea of police as crime fighters. As a consequence, their legal skills and their criminal apprehension skills have been favoured over their abilities to manage street life.

For the police, the instrumental appeal of 'broken windows' is its approach to the allocation of police resources. 'Broken windows' argues that there are two scenarios where a heavy investment in police presence – namely through foot patrols – is unlikely to be effective. The first involves communities where informal social controls are very strong. Here the need for the police to reinforce communal standards is very rare. On the other hand, there are communities that are 'unreclaimable', communities that are highly demoralized and crime-ridden. What is required of the police is for them 'to identify neighborhoods at the tipping point – where the public order is deteriorating but not unreclaimable, where the streets are used frequently but by apprehensive people, where a window is likely to be broken at any time, and must quickly be fixed if all are not to be shattered' (Wilson and Kelling 1982: 37). Drawing from Johnston (1997), such communities are imagined as 'communities of risk' as much as 'communities of collective sentiment'.

The symbolic appeal of 'broken windows' – just as we have seen with restorative justice – is its resonance with broader strands of communitarian thought. Like so many waves before it, it brings 'community' into the security debate in a way that does not supplant the 'social' as the fundamental basis of collective life. The communitarian vision of 'broken windows' has significant implications for the identification and mobilization of community capacity. As discussed earlier, RJ sees individuals and communities as active participants in the problem-solving process – they bring into the domain of policing their unique knowledge of the background and circumstances of those involved in crime. In Wilson and Kelling's view, community members play a key role in enforcing communal standards of conduct as *de facto* instruments of sovereign authority. It is because of this that Hughes (1998) describes the communitarianism of 'broken windows' as 'moral authoritarian'.

Both RJ and 'broken windows' share a conception of community as one of 'collective sentiment' (Johnston 1997). What distinguishes RJ is its desire to acknowledge diversity in the ways in which matters of security are analysed and addressed. Hughes thus sees RJ as the beginning of a more 'radical' variant of communitarianism (Hughes 1998). For Kelling and Coles crime control has been held 'hostage to social ideology' (Kelling and Coles 1996: 378). They argue that a 'root causes ideology' has led to a paralysis among practitioners because they do not possess the capacity to address the range of economic and social factors underlying criminal behaviour. At the same time, they argue that practitioners have been equally paralysed by the 'far right' who have been stressing the relationship between family breakdown and crime rates. Both of these ideological shifts have thus led to a 'de-policing' of the crime problem (Kelling and Bratton 1998: 1230).

A common variant of the 'broken windows' approach has been described as 'zero tolerance policing' (ZTP), or in some contexts, 'confident policing' (Dennis and Mallon 1997). The ZTP strand picks up the 'disorder-crime' nexus and its risk management implications. What it obscures is the image of the discretionary and friendly cop on the beat that Kelling (1999) and others (Young 1998; Goldstein 1999) argue is essential to the original Wilson and Kelling thesis. As Palmer has expressed it

> 'zero tolerance' adapts the 'broken windows' thesis in a particular manner. While both share a concern with minor disorder, the former focuses more heavily on hard-edged policing and law enforcement with an emphasis on producing results. The latter seeks to develop longer term partnerships that enhance the capacity to address underlying causes. (Palmer 1997: 234)

In essence, ZTP brings in coercive capacity as a central means of managing risk. In the name of crime prevention, ZTP privileges coercion and law enforcement as methods of 'towing the line' against disorder and civility.

New York's 'Police Strategy Number 5, Reclaiming the Public Spaces of New York' consisted of 'a full-scale effort to eliminate nuisances previously largely ignored or tolerated by beat officers' (Logan 1999: 335). This approach was drawn from Police Commissioner Bratton's experience in the city's subway system where he implemented, in Logan's words, 'an aggressive, wide-scale crackdown on turnstile-jumpers and those responsible for the rampant low-level disorder that then permeated the subways' (1999: 339). In Bratton's words,

> many of the people we were arresting were exactly the ones who were causing other problems once inside the subway system. By

> focusing on fare evasion to control disorder, we were preventing a lot of the criminal elements from getting on the trains and platforms in the first place. (Bratton and Knobler 1998, cited in Logan 1999: 340)

Bratton's strategy of targeting low-level disorder created opportunities for the police to check for criminal violations such as possession of firearms and outstanding warrants (Cuneen 1999; Logan 1999: 341).

The symbolic appeal of ZTP for police can be found in the way in which it claims to 're-police' the crime problem and 'to disprove sceptics who claimed that the police can do little about crime and disorder' (Weisburd *et al.* 2003: 425). It does so by drawing on the community symbolism of 'broken windows' to resituate the police at the centre of policing.

For Goldstein (1999), ZTP can be compatible with a problem-oriented policing approach, provided law enforcement is thought of as only one among many options. However, '[c]oupled with metaphors about war on crime, on invading neighbourhoods and about winning battles, the two words "zero tolerance" serve to assuage fears and to meet the public's primal need for attention to their security' (Goldstein 1999). In this way, Bayley (1999) argues that ZTP may indeed be a 'slogan of counter-revolution' that in our terms brings Hobbes squarely back into the picture. It does so not only by bringing coercive capacity and law enforcement back to the centre of policing, but has the added advantage of advocating 'a generic policy for the police that will work everywhere and in every place' (1999: 370).

Bratton re-imagined professional, traditional policing by addressing the kinds of bureaucratic inefficiencies that Goldstein was so critical of. He addressed the 'means over ends syndrome' by establishing the organizational structures and managerial processes to hold police managers accountable – his Compstat processes – that enabled police managers to link daily operational activities to crime reduction targets. For Bratton what ZTP does is enable police to focus on their 'core business' – the application of coercion. This is accomplished through partnerships that harness police capacities while allowing other agencies and groups to target the 'root causes' of crime that social prevention advocates. This is partnership policing that enables police to do what they are seen as doing best. In Bratton's words, what has happened in New York is 'better, smarter and more assertive policing in partnership with the criminal justice system and the community we serve' (Bratton 1998: 40). ZTP, in this view, is nodal policing that recognizes the strengths of various nodes, and networks them in ways that brings these strengths together in an effective manner. Whatever one might think of Bratton's claims – and there has been considerable scholarly debate about them – (Fagan *et al.* 1998; Bowling 1999; Greene 1999; Matthews 1992; Sampson and Raudenbush 1999; Taylor 2001) the appeal of 'broken windows', particularly in the ways in which it was realized in New York, can be explained by its alignment of seemingly contradictory elements, both radical and conservative, from previous waves in ways that established a telling political consensus. ZTP is radical in the way it took up the current trend of incorporating thinking from the business sector (Braithwaite 2003b) with a preventive, risk-based logic that refigured 'means' and 'ends' in policing (Johnston 1997, 2000a). Its conservative dimension rests in the way it brought back into policing a Hobbesian conception of the police and their role in policing.

In one of the most recent policing waves – intelligence-led policing – what has emerged is yet another integration of risk-oriented thinking that links both business and scientific sensibilities.

Policing as intelligence work

The shift towards 'intelligence-led' policing (ILP) takes up many of the ideas we have explored to this point. ILP does not re-imagine the police role so much as it re-imagines how the police can be 'smarter' in the exercise of their unique authority and capacities. Similar to 'broken windows' policing, ILP projects a clear understanding that the 'core business' of the police should be consistent with traditional policing. The shift here lies in the ways in which the police are encouraged to make use of information and communication technologies more efficiently and effectively (Tilley 2003: 321, 324). In this refiguring the 'governing through crime' approach is retained through 'a preoccupation with law-breaking and law-breakers, with crime and criminals' (2003: 321). Now, however, the aim is to ensure that this is done in more scientific, evidence-led ways.

The intelligence-led approach acknowledges the literature supporting social development approaches that highlight the wide range of structural factors associated with the 'root causes' of crime. At the same time, advocates of ILP point out that repeat offender data reveals a small percentage of individuals who are responsible for a large volume of crimes. This finding has provided the basis for arguing that targeted policing that focuses on 'at risk' individuals could make an impact on the overall crime problem (Audit Commission 1993; Barker-McCardle 2000; Heaton 2000).

Rooted in this risk-based thinking, ILP looks for patterns, both at the level of 'hot spots' and in terms of prolific offenders and groups involved in a disproportionate amount of criminal activity. Using intelligence to establish patterns of criminal conduct, '[i]ntelligence-led policing involves effectively sourcing, assembling and analysing "intelligence" about criminals and their activities better to disrupt their offending, by targeting enforcement and patrol where it can be expected to yield highest dividends' (Tilley 2003: 313).

As with other waves in policing, ILP responds to calls for economic efficiency; in the case of the United Kingdom this was a central imperative for change identified in an Audit Commission report (1993). The Commission found that police organizations were not operating in a cost-efficient manner because they were not making effective use of intelligence. The ultimate response to this report, the 'National Intelligence Model' (NIM) (National Criminal Intelligence Service 2000; Flood 2004), subsequently emerged as a

> *business model* – a means of organizing knowledge and information in such a way that the best possible decisions can be made about how to deploy resources, that actions can be co-ordinated within and between different levels of policing, and that lessons are continually learnt and fed back into the system. (John and Maguire 2003: 38–9, italics in original; see Barker-McCardle 2000)

ILP dovetails with situational crime prevention in its focus on 'risky' places – or what Ericson and Haggerty refer to as 'the geography of trouble' (Ericson and Haggerty 1997: 7) – as well as 'risky' people, data on which tends to be produced through 'crime mapping' technologies. While intelligence is used to guide the deployment of coercive capacity more effectively, it is also useful in identifying where non-police actors can play a fundamental role. As with 'broken windows' policing and its 'zero tolerance' variant, intelligence-led policing can be conceptualized within a broader framework of problem-oriented policing while at the same time contributing to the 'rationalization' (Manning 2001: 84; John and Maguire 2003) imperatives of neo-liberalism and new managerialism (Chan 2001b; Manning 2001). As Cope argues: 'Crime analysis supports problem-oriented and intelligence-led approaches by reviewing large volumes of information to identify problems, thereby enabling "tailor-made" interventions to be developed, and police activity to be targeted towards identified risks' (Cope 2003: 341). While intelligence can be used 'tactically' to provide a basis for allocating police resources in the short term (e.g. patrols), it can also be used 'strategically' to inform preventive approaches that address the distribution of problems and their causes (Smith 1998; Cope 2004; Ratcliffe 2004). In this way, ILP is seen as providing the 'evidential base' for problem-solving strategies and initiatives (Ratcliffe 2003).

At a symbolic level, ILP arguably has the potential to extend the 'moral authoritarian' strand of communitarianism that is of concern to Hughes (1998) with regard to 'broken windows' and ZTP. As with other waves in policing, there is no 'natural' way of imagining the nature of collective life. Rather, this construction tends to be shaped by other discursive and ideological waves in 'communitarianism' more generally. This is recognized by Cope (2004) in her recognition of the 'problem-oriented' potential of ILP.

Our analysis points to the ways in which policing developments work from, and with, what has gone before. Each development represents a wave that draws upon other waves in ways that create new combinations of elements. New mentalities, technologies and practices are crafted from crucibles of past experience. Waves require both innovation and 'bite' for them to take hold. Bite requires resonance with established ways of understanding and with institutional concerns and objectives. Institutions matter, but so do mentalities and the innovative capacities of human agency.

From a waves-based perspective, there is no inevitable trajectory to the kinds of developments we have discussed. Whether or not the intelligence-led agenda is prompting a return to traditional policing with its own distinct solution to the 'means-over-ends' syndrome depends in part on an unpredictable and contingent mix of social, political and economic factors that influence the shape and magnitude of waves. While policing change, like all change, is contingent, the dominant pattern created by the successive waves of change that have constituted policing is that it is moving in a nodal direction. While the police continue to be central to policing, they now are seen, and see themselves, as one node among others. Within this nodal context they have sought, with considerable ingenuity, to establish a role for themselves that recognizes and builds on the fact that they have been, and will likely continue to be, a fundamentally Hobbesian institution of governance.

The most recent iteration of this wave-like process has been the emergence of what in Britain is called 'reassurance policing'. Like 'broken windows' policing it seeks to reassure citizens that security is indeed being governed (Crawford and Lister 2004; Innes 2004; Crawford *et al.* 2005; Millie and Herrington 2005; Herrington and Millie 2006; Fielding and Innes 2006; Blair in press). What this wave rediscovers is that it is not simply order at any particular moment that citizens value, but the assurance that order will be maintained over time (Innes 2004). Thomas Hobbes expressed this centuries ago when he compared peace to good weather. Good weather, he argued, did not refer simply to a moment of sunshine, but to an assurance that sunshine can be expected to continue.

Policing as reassurance

In the United Kingdom there is now an explicit public policy agenda to provide public police services that are not only effective but are also responsive to the 'reassurance gap' (Millie and Herrington 2005) between citizens' fear of crime and actual rates of crime (Blair 1998; 2003). Emblematic of this wave of reassurance policing has been the introduction by Sir Ian Blair, Commissioner of the London Metropolitan Police (the Met), of community support officers (CSOs) (Blair in press). CSOs represent a second tier of policing service within the Met. They are 'owned' and trained by the Met but possess fewer legal powers than a regular police constable. Blair writes:

> The simple great advantage of CSOs is that they do not leave the street. They do not need extensive training, they do not make arrests, they do not police public order or football matches, they do not go to court, they just are. They provide the presence which we believe the public are seeking and which will stem incivilities. (Blair in press)

As this insightful comment recognizes, the reassurance policing wave extends the broken windows wave by deliberately placing CSOs in areas where levels of incivility are linked to heightened levels of public fear and anxiety.

The establishment of the CSOs also represents an extension of the intelligence-led wave. Blair argues that the mandate of the public police has widened. Not only must they place greater emphasis on disorderly behaviour, they must also enhance their capability to address issues of serious crime and terrorism (see Chapter 3), particularly in the wake of the London bombings of July 2005 (Blair 2005). The public policing enterprise, according to Blair, must be organizationally embedded in local communities in order to most effectively identify, prevent and respond to the range of behaviours that exist across a broad spectrum ranging from minor disorder to more serious crime. As Blair notes:

> [N]ational security depends on neighbourhood security.
> It will not be a Special Branch officer at Scotland Yard who first confronts a terrorist, but a local cop or a local community support officer.
> It is not the police and the intelligence agencies who will defeat crime and terror and anti-social behaviour; it is communities.
> We do not want one kind of police force being nice to people and another one arriving in darkened vans wearing the balaclavas.
> Whoever is responsible for the one has to be responsible for the other. (Blair 2005)

For Blair it was their patrol function that brought police into communities and enabled them to work with communities to establish nodal partnerships. As this core feature of policing has declined, he argues, so too have the close ties between citizens and communities that are so essential to effective nodal policing. This decline, Blair argues, should be reversed if the balkanization of policing that it has engendered – as people have sought to fill the policing gap that the decline of patrol has left vacant – is to be challenged. Such a reversal, he argues, is essential if the police (and by implication states) are to reclaim their Hobbesian status as security guarantors. To accept balkanization, he argues, would be to return to the plural features of policing that the British police were established to bring to an end. In a reference to Peel, Blair argues that if police do not once again engage in patrol as an essential part of their role we might find them embarking on 'a return to the policing arrangements of pre 1829' (Blair 2003: 6). He states 'Unless we do something dramatic, we are sure that public dissatisfaction will grow and that local authorities, housing associations and private individuals will cease to look to the Met for a solution' (2003: 6). The emergence of reassurance policing in the United Kingdom has been studied by Crawford and colleagues (Crawford and Lister 2004; Crawford *et al.* 2005). They provide compelling evidence in support of Blair's position:

> Visible patrolling now constitutes a central element of an emerging market among an array of purchasers and providers in diverse locations – residential, industrial and commercial

> areas. This new market has provided residential communities, social housing providers and commercial businesses with new choices and opportunities concerning the provision of security. Consequently, the public police are becoming part of a more varied and complex assortment of organisations and agencies with different policing functions together with a more diffuse array of processes of control and regulation. In this context, rather than a monopoly of control by the police we stand on the brink of a more complex future in which alliances of public, parochial and private agencies and interests are drawn together in intricate networks of policing. (Crawford and Lister 2004: 414)

In this commercial environment the hope that lies behind Blair's initiative is that the Met will re-emerge within a nodal context as the pre-eminent supplier within a reassurance policing market (*Economist* 2005).

This strategic positioning is essential, according to Blair, for both instrumental and symbolic reasons. His vision for an 'extended police family' (*Economist* 2005; Johnston 2003) is one of effective and high-quality intelligence-gathering that is housed and managed within a 'single police service' devoted to engagement with and accountability to communities (Blair 2005). It is a vision that may reflect, as Johnston suggests, 'the latest attempt to secure police sovereignty over the governance of community policing/safety' (Johnston 2003: 188) within a world of nodal governance.

As we observe this latest wave, we are reminded that transformations in public policing are driven not simply by the strategic and normative aspirations of police managers and the authorities to whom they are accountable. Police reform is, for the most part, concerned with improving the way police organizations do business, not only in instrumental terms, but also in terms of deepening their commitment to democratic values. At same time, however, police leaders are cognizant that they now operate within a world of nodal contests. As a consequence they have become increasingly concerned with developing strategies that will enable them to effectively jockey for positions to ensure that police institutions continue to play a central role in the governance of security (Dupont 2006a).

In the market for reassurance policing the public police are no longer able to take for granted their strategic and symbolic position. This is not to suggest that the police no longer possess considerable symbolic power (see Loader 1997), nor that they cannot lay claim to a unique set of knowledge, expertise and capacity. It is, however, to suggest that future waves in governance may be increasingly shaped by a diverse array of police and non-police actors involved in imagining, directing and delivering local security goods.

Acknowledging the nodal reality of security governance is important not only for police organizations, who may be required to think more explicitly in terms of how they position themselves *vis-à-vis* other nodes, particularly commercial nodes, in the security field (Wood 2006a). This reality also provides opportunities for other nodes in civil society – including those who exist outside the public and private sector – who do, or could, bring to bear significant forms of knowledge and capacity to the governance of security.

A significant theme running through our analysis of waves of change concerns the central role that police organizations have played in the establishment of partnerships and networks. In all of the waves we have discussed, police organizations have recognized the importance of diverse forms of knowledge and capacity. However, the police have invariably used this recognition to privilege the police role as the pre-eminent policing node. This police-centred view of nodal partnerships, while both predictable and understandable, has limited what is thinkable and this in turn has limited innovation (Wood and Marks 2006). We respond to this in Chapter 4 where we consider the possibility of creating 'design principles' that will enhance the ability of 'weak actors' to play a more central role in nodal governance.

Conclusion

In this chapter we have explored shifting imaginings of community security within the field of public policing. Our analysis focused on how old and new ways of thinking mix and meld to produce 'waves' of change. We have focused in particular on shifting mentalities of punishment and of risk and their implications for the forms of knowledge and capacity that are seen as essential to the local governance of security. We have traced how the knowledge and capacity of the police has been rethought and re-articulated in successive waves of change, and examined how the police role has been refigured in ways that imbricate police within nodal relationships.

We deployed the imagery of waves as a metaphor to help us recognize the fluidity of change. The developments we have discussed have not unfolded in a linear fashion. Instead they have displayed a back and forth character that has brought the old and the new together in a variety of permutations. These combinations have not always been coordinated and orchestrated but have involved contest and tension. The future of public policing will most likely continue to be expressed through frequently uneasy combinations of mentalities, institutional arrangements and technologies. Some waves may dominate others in the future, or may lead to new waves that we do not yet anticipate.

In the next chapter we continue to explore this theme of contest and coexistence. At the global level, as with the local level, there have been a series of transformations in the governance of security that speak to the nodal nature of the world in which we live. We now move beyond imaginings of 'community security' to those of 'human security' in exploring other mentalities and strategies of governance. These imaginings imply a significant role for new nodes and nodal assemblages of governance and as such provide new ideas for what is possible.

Chapter 3

Human security and global governance

Introduction

In the previous chapter we explored shifting imaginings of community security within the field of public policing. We examined the mixing and melding of old and new ways of thinking within successive 'waves' of change. We focused in particular on punitive and risk-oriented mentalities associated with ongoing efforts to rethink the place and role of the public police in relation to other nodes. Community policing, as a state-led agenda, has involved police organizations playing an active role in constituting and steering nodal relationships. This agenda has been and continues to be shaped by the instrumental and symbolic objectives of police managers and the state institutions and political constituencies to whom they are accountable.

In this chapter we move our focus to the global realm by exploring imaginings of human security. The conceptual move from states, and even communities, to *human beings* as 'referent objects' (Buzan *et al.* 1998) is significant for several connected reasons. It represents a fundamental challenge to traditional understandings of just what 'security' is, how it can be promoted and through what sources of knowledge and capacity. Within a state-centred paradigm, the territorial integrity of nation states is paramount, as is the military capacity of states to protect this integrity. Military and police organizations, as international and domestic specialists in the use of force, are therefore seen as paramount in the governance of security. A human security approach seeks to decentre the state as referent object while examining security threats that arise from nodes and networks of people and activities that defy traditional state boundaries. From this standpoint, coercive capacity exercised by or on behalf of states is simply one among a plethora of capacities required in securing people.

As with other developments in community security governance, it is not the case that the dream of human security governance reflects an abandonment of other dreams. Rather, state-centric, community-centric and human-centric imaginings and practices coexist; they mix and meld. In some cases – particularly in respect to the 'war on terror' – this mixing and melding can be fraught with tensions, as some of the objectives underlying state-centric governing practices may collide rather than collude with human-centric initiatives. The question of just precisely what dreams of governance may 'win out' in the end, or dominate the field of security writ large, may simply reflect an imbalance of power, in the Latourian sense, whereby some actors/nodes have greater capacity to imagine security in particular kinds of ways and to enrol other nodes and networks towards their stated objectives. Indeed, it could be said, just as Braithwaite and Drahos (2000) have done, that the difference between 'weak actors' and 'strong actors' rests in the capacity to ensure that some dreams of governance are more effectively realized than others. In this chapter we examine ways in which human security is imagined and the mentalities, institutions and practices of global governance supported by this imagining.

Imagining human security

In the traditional Realist school of International Relations (IR), the state is the primary referent object, whereby security is imagined in terms of the (sovereign and territorial) integrity of the nation-state (Møller 2000). Put simply, the state is the 'principal actor' in the Westphalian system that frames Realist thinking (2000: 44). From a human security perspective, this state-centred view is displaced by a much broader conceptualization of the security of people. The human being is now the central referent object, whereby 'people's interests or the interests of humanity, as a collective, become the focus' (Commission on Human Security 2003: 3). This conceptual move is advanced in the United Nations Human Development Report of 1994:

> For too long, the concept of security has been shaped by the potential for conflict between states. For too long, security has been equated with the threats to a country's borders. For too long, nations have sought arms to protect their security.
>
> For most people today, a feeling of insecurity arises more from worries about daily life than from the dread of a cataclysmic world event. Job security, income security, health security, environmental security, security from crime – these are the emerging concerns of human security all over the world.
>
> This should not surprise us. The founders of the United Nations had always given equal importance to people's security and to territorial security. As far back as June 1945, the US secretary of state reported this to his government on the results of the San Francisco Conference:
>
> *The battle of people has to be fought on two fronts. The first is the security front where victory spells freedom from fear. The second is the economic and social front where victory means freedom from want. Only victory on both fronts can assure the world of an enduring peace… No provisions that can be written into the Charter will enable the Security Council to make the world secure from war if men and women have no security in their homes and their jobs.* (UNDP 1994: 3, italics in original)

This human-centric vision expresses what contemporary scholars like Held (2004) and Kaldor (1999) describe as a 'cosmopolitan' ethic which imagines 'human beings living in a world of human beings and only incidentally members of polities' (Barry 1999: 35 cited in Held 2003: 469). As Held describes it, 'the ultimate units of moral concern are individual people, not states or other particular forms of human association. Humankind belongs to a single moral realm in which each person is equally worthy of respect and consideration' (2003: 470).

While this human-centric focus captures the heart of the cosmopolitan vision, the Commission on Human Security (hereafter referred to as 'the Commission') clearly points out that this focus can be, and ought to be, aligned with a politics of state security. On this point, Kerr (2003) observes an 'evolving dialectic' between 'state-centric' and 'human-centric' conceptions. At the same time, it is recognized that some states have violated, and continue to violate, the security of their own people:

> This understanding of human security does not replace the security of the state with the security of people. It sees the two aspects as mutually dependent. Security between states remains a necessary condition for the security of people, but national security is not sufficient to guarantee people's security. For that, the state must provide various protections to its citizens. But individuals also require protection from the arbitrary power of the state, through the rule of law and emphasis on civil and political rights as well as socio-economic rights. (Frene Ginwala cited in Commission on Human Security 2003: 3)

The threats that states themselves pose to the security of people are, of course, only part of a broader array of threats that derive from a variety of sources. In the following section we examine some key threats, particularly those that are a feature of our current era of globalization.

Threats to human security

The twin pillars of 'freedom from fear' and 'freedom from want' expressed in the above text from the United Nations denote a vast range of social, political, economic and cultural factors that may threaten the security of human beings. While crime and violence represent significant threats to human security, an array of other issues – previously conceptualized in non-security terms – are 'securitized' (Buzan *et al.* 1998). According to Buzan, Wæver and de Wilde, 'securitization' is a process of social construction involving those who carry out the speech act ('securitizing actors') who articulate an existential threat to a referent object. In constituting something as a security problem, one is calling for emergency or extraordinary measures to deal with it (1998: 23–4). 'Security' is thus not an objective state of affairs, but a 'speech act' (Wæver 1995; Buzan *et al.* 1998); '[it] is not of interest as a sign that refers to something more real; the utterance *itself* is the act' (Wæver 1995: 55). In a similar vein, Duffield and Waddell suggest that the concept of human security should be understood as a 'principle of formation. That is, as producing the "humans" requiring securing and, at the same time, calling forth the state-non-state networks of aid, subjectivity and political practice necessary for that undertaking' (Duffield and Waddell 2006: 2).

Claims that human security is threatened by crime, violence and war are currently made by an array of securitizing actors from within and across state and non-state sectors. Such actors are concerned in particular with the increasingly transnational character of what Raab and Milwaard (2003) describe as 'dark networks' including organized crime (e.g. human trafficking, money laundering, drug smuggling) and terrorism (Commission on Human Security 2003).

Research has revealed the highly nodal and networked character of the actors and activities involved in the commission of organized crime (Arquilla 2001; Paoli 2002; Beare 2003). Transnational organized crime is carried out by sophisticated and agile nodes that have had success in escaping the gaze of law enforcement agencies at local, national and international levels (Wood 2006b). As Dupont (2006b) describes it, 'the myth of the fight against crime and the war-like metaphors on which it relied were seriously undermined by the discovery that organized crime was much less organized than initially thought and that crime syndicates were loosely coupled alliances of individuals who retained a large degree of autonomy'. Even more, such loosely coupled alliances can be forged across the imagined boundaries of licit and illicit markets (Beare 2003). In some cases, collusion between illicit market actors and corrupt government actors has been discovered (Ruggiero 2003), making it all the more challenging to separate out legitimate from illegitimate forms of conduct (Beare 2003).

A similar nodal analysis has been used in examining the new face of terrorism. According to Gross Stein, transnational terrorist nodes are assembled in networks that create a kind of organizational redundancy, whereby the removal or destruction of one node does not affect the resiliency of the network as a whole (Gross Stein 2001: 74). The violence associated with acts of crime and terror is indicative of the changing nature of violence and war more generally in the wake of globalization. Contemporary violent conflict poses a particular set of threats to human security (Commission on Human Security 2003; Human Security Centre 2005) in contrast simply to those posed by battles between states or what Kaldor (1999) describes as the 'old wars'. Kaldor argues rather that the object of our analysis should be the 'network warfare' that is carried out by 'armed networks of state and non-state actors' (Kaldor 2003: 119) of which there is a vast array: 'paramilitary groups organized around charismatic leaders, warlords who control particular areas, terrorist cells, fanatic volunteers like the Mujahadeen, organized criminal groups, units of regular forces or other security services, as well as mercenaries and private military companies' (2003: 119).

In contrast to the geo-political or ideological nature of the old wars, the 'new wars' (Kaldor 1999; Duffield 2001) express an identity politics involving a clash between 'particularistic identities' and what Kaldor describes as a 'cosmopolitanism, based on inclusive, universalist, multicultural values' (1999: 6; see also Commission on Human Security 2003: ch. 2). It is indeed the case, Kaldor argues, that old wars were concerned with identity politics as well, but it was a kind of politics centred on 'a notion of state interest or to some forward-looking project – ideas about how society should be organized' (1999: 6). The new wars, which centre on identities related to nationhood, clans, religion or language, embody nostalgia for traditional social cleavages that were tempered through colonialism and/or the Cold War. Such identities are being re-imagined and redeployed in the midst of a failure in, or decline in legitimacy of, previous political projects such as socialism or the nation-building agendas of post-colonial leaders (1999: 6–7). The 'backward-looking political projects' of the new wars thus 'arise in the vacuum created by the absence of forward-looking projects' (1999: 7). Kaldor adds: 'Unlike the politics of ideas which are open to all and therefore tend to be integrative, this type of identity politics is inherently exclusive and therefore tends to fragmentation' (1999: 7).

A second core feature of the new wars pertains to the methods of violence used. While old wars involved the use of military might to capture territory, the new wars draw inspiration from guerrilla warfare and counterinsurgency movements while adding a different dimension. Similar to guerrilla warfare, those involved in the new wars seek to gain political control of populations while avoiding military battles as much as possible. In contrast to guerrilla warfare which was devoted, in theory, to winning the 'hearts and minds' of populations, the new wars seek to destabilize – as in counterinsurgency (Kaldor 1999: 8) – for, as Braithwaite (2006) describes it, 'hatred and fear across the schisms of dominated populations are necessary to keep these wars going'. Kaldor explains:

> The aim is to control the population by getting rid of everyone of a different identity (and indeed of a different opinion). Hence the strategic goal of these wars is population expulsion through various means such as mass killing, forcible resettlement, as well as a range of political, psychological and economic techniques of intimidation. This is why, in all these wars, there has been a dramatic increase in the number of refugees and displaced

> persons, and why most violence is directed against civilians. (Kaldor 1999: 8)

The organizations that carry out the new wars are decentralized rather than hierarchical and consist of a variety of nodes, including 'paramilitary units, local warlords, criminal gangs, police forces, mercenary groups and also regular armies including breakaway units of regular armies' (1999: 8). These nodes take advantage of new types of weapons such as undetectable landmines and small and light firearms, as well as communication technologies like mobile phones and computer links that allow for coordination of activities (1999: 8). Globalization has allowed the identity politics of the new wars to be played out on both local and global stages. Information technology, combined with the ease of world travel, has enabled diaspora communities, for example, to carry out their political projects as global networks (1999: 7).

The third core feature of the new wars is the 'globalized war economy' that supports them. While old wars were sponsored by states, the new war economy is decentralized (Kaldor 1999; Duffield 2002). Weak and failing states provide a 'fertile environment' for nodes and networks to prosper. While possessing some 'trappings of statehood', weak states lack public legitimacy and are increasingly deficient in their control over the exercise of legitimate violence (Kaldor 2003: 120). Violent actors are highly adaptive to the weakening economic bases of states by seeking innovative forms of financial sponsorship through a new kind of global formal economy (2003: 120–2; Braithwaite 2006). Duffield elaborates:

> War economies ... have linked local resources, such as alluvial diamonds and tropical hardwoods, or the derivatives of coca and poppy production, both illegally and legally to global markets. They have also established transborder nodal connections with the grey world of the arms trade, money laundering and international criminal syndicates. Besides illegal trade, warring parties and terrorist groups have also established legal businesses. In Africa the al-Qaida network, for example, has run companies operating in the fields of import-export, currency trading, civil engineering and agriculture and fisheries. (Duffield 2002: 157)

Given the nature of such transborder nodal connections, Kaldor notes that while she uses the term 'war' to reflect the political character of violence, she states that the new wars effectively blur the distinctions between the established categories of traditional war, organized crime and mass violations of human rights (1999: 2).

New forms of data are being collected on both the direct and indirect costs of the new wars. A recent report by the Human Security Centre (2005) states that 'the overwhelming majority of today's armed conflicts are fought within, not between, states and that most take place in the poorest parts of the world' (2005: 15). In fact the Centre's data shows that '95 per cent of armed conflicts have taken place within states, not between them' (2005: 18). That being said, Kaldor emphasizes the global dimension of localized violence. She suggests that while some of the new wars are often described as 'civil wars' or 'low-intensity conflict', and are local in their manifestations, 'they involve a myriad of transnational connections so that the distinction between internal and external, between aggression (attacks from abroad) and repression (attacks from inside the country), or even between local and global, are difficult to sustain' (Kaldor 1999: 2).

A dataset created jointly by the Uppsala University's Conflict Data Program and the International Peace Research Institute, Oslo (PRIO) covers the period from 1946 to 2003 and is based on four different types of conflict. 'Interstate' conflict refers to conflict between states while 'intrastate' pertains to conflicts within states. The third category of 'extrastate' conflict refers to 'a conflict between a state and a non-state group outside of the state's own territory' and 'applies primarily to wars fought to gain independence from colonial rule' (Human Security Centre 2005: 20). The fourth category of 'internationalised internal conflict' refers to 'intrastate conflict in which the government, the opposition, or both, receive military support from another government or governments, and where the foreign troops actively participate in the conflict' (2005: 20). Uppsala also collected data for 2002 and 2003 on two other types of conflicts that had previously been excluded from conflict datasets. The first type is that which occurs between non-state actors, including conflicts between warlords or between religious or ethnic groups (2005: 21). The second type is what Uppsala describes as 'one-sided' violence, including genocides and massacres (2005: 21).

While traditional wars were carried out by huge armies deploying conventional weapons, which resulted in high rates of battle deaths, the new wars are either 'low-intensity' or 'asymmetric' in nature and take place largely in developing countries. Low-intensity civil wars are often carried out by 'relatively small, ill-trained, lightly armed forces that avoid major military engagements but frequently target civilians' (2005: 34). Asymmetric conflicts, as seen currently in Iraq and Afghanistan, 'involve US-led "coalitions of the willing", using high-tech weaponry against far weaker opponents who have few or no allies' (2005: 34).

Transformations in warfare can also be understood in reference to three key shifts in the nature and organization of military functions – the deployment of child soldiers; the use of paramilitary forces; and outsourcing to private military firms (2005: 34). The use of child soldiers is said to be growing. While recruitment of young people under the age of 18 is prohibited by the UN Convention on the Rights of the Child (Optional Protocol on the Involvement of Children in Armed Conflicts 2000) (2005: 35), '[c]hildren fight in almost 75 per cent of today's armed conflicts' (2005: 35). Children are also recruited for terrorist missions in Northern Ireland, Columbia and elsewhere. The rise in recruitment of child soldiers is partly explained by the availability of cheap, lightweight weapons that do not require the physical strength of an adult. In addition, rates of youth unemployment provide 'a pool of potential recruits who may have few other survival options' (2005: 35).

Similar to the factors that explain the rise in numbers of child soldiers, the growth in paramilitary organizations can be linked to the availability of inexpensive arms as well as to the fact that they can be trained quickly and require low levels of logistical support (2005: 35). Paramilitaries include a range of entities, from armed police to riot squads to intelligence agencies to private armies, and carry out different functions such as protecting established regimes from internal threats like separatist rebellions or military coups (2005: 35, 37). They can also undertake riot control and border security and even participate in the 'elimination of political opponents' (2005: 37).

Warfare has also been transformed through the use of private military service providers (Mandel 2002; Singer 2003; Avant 2005). This has been triggered in part by a steep decline in both defence spending and military aid to developing countries following the Cold War (Human Security Centre 2005). Private military firms (PMFs) offer a variety of services, including maintenance and infrastructure, logistical support, strategic advice and front-line combat (Singer 2001, 2003). The 'market for force' (Avant 2005) is such that '[h]undreds of PMFs have operated in more than 50 countries, and their global revenue has been estimated to exceed US$100 billion a year' (Human Security Centre 2005: 37).

As we discuss in Chapter 5, this highly nodal character of security provision has profound regulatory implications, for it reveals the limitations of a state-centred approach devoted largely to rendering state nodes accountable to particular standards of conduct (Lewis and Wood 2006). As part of this regulatory problem, accurate assessments of the harms caused by violence that is global and nodal in character continue to be precluded by state-centred understandings that focus on indicators like numbers of battle deaths (Human Security Centre 2005). As a corrective to this, work is being done on measuring both the direct and indirect costs of violence carried out by state and non-state nodes. One such cost is the weakening or destruction of economic, political and social livelihood systems that are central to human resilience and prosperity (Human Security Centre 2005; Duffield and Waddell 2006: 6). For example, in addition to numbers of 'indirect' or 'excess' deaths, there can also be a loss in *'healthy* years of life' as a result of death or disease or other impacts of war (Human Security Centre 2005: 125–6). Furthermore, conflict and war produce social impacts such as 'damage to 'societal networks'', 'population dislocations' and 'diminished quality of life' (2005: 127). According to the *Human Security Report 2005*:

> Because many of these indirect effects may take years to manifest and are difficult to distinguish from the effects of diseases and conditions not attributable to war, they are often ignored in favour of immediate body counts. But disregarding indirect mortality and morbidity grossly underestimates both the human costs of war and the level of expenditure and effort needed to mitigate post-conflict suffering. (cited in Human Security Centre 2005: 129)

This emphasis on the indirect costs of war reflects a broader concern of the human security approach with non-military threats to people in zones of peace as well as in zones of war. A 'sustainable human development' perspective (see Hampson *et al.* 2002: 26–7) focuses on threats such as 'human-induced' problems that reveal deep inequities in the distribution of resources that are necessary for human flourishing and survival. UN Secretary-General Kofi Annan articulates this emphasis on the structures of inequity:

> Human security in its broadest sense embraces far more than the absence of violent conflict. It encompasses human rights, good governance, access to education and health care and ensuring that each individual has opportunities and choices to fulfil his or her own potential. Every step in this direction is also a step

> towards reducing poverty, achieving economic growth and preventing conflict. (cited in Commission on Human Security 2003: 4)

The human security vision – particularly its developmental strand – is thus focused on a securitization of problems previously conceptualized in non-security terms in areas such as health, the economy and sustainable development more generally. With respect to nutrition, for example, the consensus reached at the 1996 World Food Summit was: 'Food security exists when all people, at all times, have physical and economic access to sufficient, safe and nutritious food to meet their dietary needs and food preferences for an active and healthy lifestyle' (cited in AusAID 2004: 6). The related problem of 'income security' is expressed by the Commission:

> A fifth of the world's people – 1.2 billion – experience severe income poverty and live on less than $1 a day, nearly two-thirds of them in Asia and a quarter in Africa. Another 1.6 billion live on less than $2 a day. Together, 2.8 billion of the world's people live in a chronic state of poverty and daily insecurity, a number that has not changed much since 1990. About 800 million people in the developing world and 24 million in developed and transition economies do not have enough to eat. (Commission on Human Security 2003: 73)

Threats to economic security can be exacerbated by global economic processes hitting already underprivileged populations the hardest. The Commission contends that while markets and trade are necessary for economic growth, which itself is central in promoting human security, markets come with risks that can adversely affect different populations in particular ways. Individuals that are poor tend to have no security blanket, especially in developing countries where financial downturns are more severe and frequent than in developed countries. The security of people who are already poor before a financial crisis or collapse is further compromised due to a lack of subsistence income, a lack of social insurance for the unemployed or self-employed, a lack of employee benefits paid out by employers, a lack of personal savings, and finally, a lack of access to credit schemes or private insurance (2003: 75–83).

Another risk of markets is 'financial contagion', whereby economic crises in one country can 'reverberate' to other parts of world, such as when Thailand's economic crisis affected East Asia generally, as well as Africa, Latin America, Central and Eastern Europe and Russia (2003: 83). Natural disasters, including earthquakes and floods, are a key threat to markets and hence to human security. Natural disasters cause death, drought and famine, and can destroy homes and damage natural resources (2003: 82–5). It has been further suggested that concerns for human security cannot override concerns for the security of the environment and the ecosystem (Khagram *et al.* 2003). The idea of an 'intimate coupling of nature and society' (2003: 289) means that 'efforts to protect nature will fail unless they simultaneously advance the cause of human betterment; efforts to better the lives of people will fail if they fail to conserve, if not enhance, essential resources and life support systems' (2003: 289).

The two conceptual moves of the shift to human beings as the referent object and the securitization of non-military threats to people have served to open up a new space for re-imagining the ways in which security is, and could be, governed. In this new security field, those forms of knowledge and capacity traditionally devoted to the governance of state security – i.e. coercive capacity including military might – is decentred. Space is made for other sources of knowledge and capacity to be identified, developed and deployed in producing arrangements for global governance that serve to form bonds or links between otherwise distinct problem spaces.

In the next section we reflect on strategies of governance that are supported by a human security orientation. We do not intend to cover the gamut of substantive areas that form part of human security governance. Rather, we focus our discussion on ways in which policy makers, practitioners and scholars working in the human security field think more broadly about where governing capacity resides in terms of levels or horizontal layers (i.e. micro/local, national, global) and how these layers are, or should be, coordinated or integrated. We examine in particular the cosmopolitan prescriptions of scholars like Kaldor and Braithwaite who stress the importance of privileging forms of local knowledge and capacity within a global framework of norms.

Strategies of human security governance

The most obvious implication of the human security approach is what it requires in terms of identifying, mobilizing and in some cases building diverse forms of knowledge and capacity that can be harnessed at local, national and global levels. Given the need for multiple auspices and providers according to different mentalities, one might describe the promotion of human security as polycentric. McGinnis defines this concept as follows:

> A political order is polycentric when there exist many overlapping arenas (or centers) of authority and responsibility. These arenas exist at all scales, from local community groups to national governments to the information arrangements for governance at the global level ... A sharp contrast is drawn against the standard view of sovereignty as connoting a single source of political power and authority that has exclusive responsibility for determining public policy. (McGinnis 1999a: 2)

The overlapping centres of authority and responsibility that are or could be involved in the governance of human security cover the gamut of strategies for addressing issues of 'freedom from fear' (especially crime and terrorism) as well as issues of 'freedom from want'. In the following subsections we highlight some key governance modes or forms, and the types of knowledge and capacity they entail for their actualization (see Hampston *et al.* 2002 generally on approaches to human security).

Fighting crime and terror

The current emphasis on global strategies for fighting crime and waging 'war' on terrorism is something to which we return towards the end of this chapter as we discuss the challenge of aligning state-centric and human-centric approaches. For the moment we focus on the fact that the transnational character of crime and terrorism has called for innovative responses on the part of traditional institutions of security governance, particularly in the criminal justice, intelligence and military sectors. Such responses have required, and will require, new types of nodal arrangements for identifying, harnessing and coordinating sources of knowledge and capacity that can contribute to the prevention of criminal and terrorist activities. As reflected in the previous chapter, the idea that police organizations, for example, ought to develop local partnerships or networks is not wholly new; in fact it has been under development for over two decades in most western democracies (Fleming and Wood 2006).

Policy makers and practitioners espouse the virtues of partnerships and networks in relation to their potential to enhance both reactive and proactive policing. An understanding of the 'dark network' character of transnational activities like organized crime and terrorism now explicitly informs the strategic directions of police agencies, military organizations and other state security institutions. In Australia, for example, the new Organized Crime strategy of Victoria Police is based on the assumption that its organization must 'change some of its internal structures, culture and thinking in order to succeed in matching and combating the fluid, flexible, dynamic and networked characteristics of organized crime networks' (Victoria Police 2005: 8). In a similar vein, the Australian Federal Police is developing its international capacity to be 'intelligence-led' through the formation of global networks consisting of public agencies, communities, the private sector and international organizations (Wardlaw and Boughton 2006).

Global networks of policing and security also form a central plank of new anti-terrorism policy. State governments have set out to develop in general terms what might be described as knowledge networks, capacity networks and resource networks (Wood 2006b). With respect to knowledge networks, for example, state agencies recognize the importance of 'collecting' and 'connecting the dots' (RAND Corporation 2004) based on the assumption that 'forestalling major threats such as terrorist attacks or epidemics requires weaving together disconnected pieces of information to reveal broader patterns' (2004: xi). The 'war on terror' involves the use of military strategies as well as other strategies of intelligence and surveillance designed to track the flows (including funds, information and people) of dark networks. Such strategies have required the formation of cooperative arrangements in areas such as the sharing of intelligence and are aimed at preventing terrorist attacks by hindering the flows of resources required to carry out terrorist activities and apprehending perpetrators (Commission on Human Security 2003: 23; see also Duffield and Waddell 2006; Stohl and Stohl 2002, 2004).

Capacity networks are similarly being formed on the assumption that specialist agencies like the police do not possess all of the skills required in identifying, retrieving and analysing pieces of information that may assist in crime prevention or crime control. The increasing use of civilians, with specialist analytical capacities, is one example of police and other state agencies establishing more robust webs of capacity. Resource networks are formed in a variety of ways. Ayling, Grabosky and Shearing (2006) argue that state institutions like the police can 'pull in' a variety of external resources through 'coercion' (e.g. requiring certain organizations like chemical companies to report suspicious sales), 'sale' (e.g. hiring private military contractors) and 'gift' (e.g. receiving donations of equipment) (Ayling and Grabosky 2006).

In broad terms, national governments are seeking to reinvent themselves in ways that respond more effectively to the nature of the 'new wars'. The 9/11 Commission Report states:

> As presently configured, the national security institutions of the US government are still the institutions constructed to win the Cold War ... Instead of facing a few very dangerous adversaries, the United States confronts a number of less visible challenges that surpass the boundaries of traditional nation-states and call for quick, imaginative, and agile responses. (National Commission on the Terrorist Attacks upon the United States 2004: 399)

Central to this agility is the capacity to enlist domestic and international organizations in supplying knowledge, capacity and resources within an overall structure that achieves 'unity of effort' in areas such as intelligence gathering and operational planning (2004: ch. 13).

Protecting people in zones of conflict

As with efforts to fight crime and terror, the protection of people in zones of conflict requires the involvement of traditional state capacity in the form of police, military and peacemaking efforts. Yet, this protection function as it has been conceptualized by the Commission on Human Security within an 'integrated human security framework' is something that should be built upon five pillars that require multiple sets of resources: 'ensuring public safety'; 'meeting immediate humanitarian needs'; 'launching rehabilitation and reconstruction'; 'emphasizing reconciliation and coexistence'; 'promoting governance and empowerment' (Commission on Human Security 2003: 61). 'To the extent possible', the Commission writes, 'all relevant tools and instruments – political, military, humanitarian and developmental – should come under unified leadership, with integration close to the delivery points of assistance' (2003: 61).

The Commission notes that states rarely have the capacity to deliver safety in their own post-conflict contexts. In such situations, forms of violence and crime often increase, and in some cases state institutions of security – particularly the police and military – are corrupt. While outside military intervention is required for various stabilization roles including separating combatants and protecting the provision of humanitarian and reconstruction assistance, 'the goals should shift towards upholding public safety through fighting crime (domestic and transnational) and building the capacity of national and local police' (2003: 62). The Commission adds: 'Effective state security institutions upholding the rule of law and human rights are an essential component for achieving human security, development and governance' (2003: 63). In a recent piece on international networks of security governance, the Commissioner of the Australian Federal Police discusses the virtues of capacity-building and knowledge transfer in weak and failing states of the Asia-Pacific region (Keelty 2006). While such capacity-building is partly guided by the desire to enhance unity of effort across international police organizations in global law enforcement, it is also guided by a broader cosmopolitan ethic, as we discuss further below, that is concerned with promoting state compliance with human rights and other global norms.

'Peacemaking' has traditionally been understood as an enterprise devoted to overseeing the implementation of an agreement reached between warring parties. Kaldor challenges this understanding. She argues that what is required rather is the 'enforcement of cosmopolitan norms, i.e. enforcement of international humanitarian and human rights law', and adds that 'it ought to be possible to devise strategies for the protection of civilians and the capture of war criminals. The political aim is to provide secure areas in which alternative forms of inclusive politics can emerge' (Kaldor 1999: 124–5). The cosmopolitan law enforcement officer that Kaldor imagines would represent a hybrid of soldiering and policing roles. While tasks may entail the control of airspace or the maintenance of ceasefires, as traditionally carried out by soldiers, other functions – such as protecting freedom of movement and capturing war criminals – come closer to the policing function (1999: 125).

Kaldor's vision of cosmopolitan law enforcement espouses the principle of 'impartiality' rather than 'neutrality'. While not discriminating on the basis of race, religion or other forms of identity, it would be the job of the law enforcer to protect those threats to human security that are enshrined in cosmopolitan norms (1999: 128). Furthermore, no nation should be immune from such norms, including the United States (Kaldor 2003: 156). The International Criminal Court is also central to Kaldor's vision of a multilateral architecture of cosmopolitan law enforcement (2003: 156) as well as other structures, resources and capacity-building required to support 'a new kind of soldier-cum-policeman' (Kaldor 1999: 130) representing a 'professional service, which would include both civilian and military personnel, ranging from robust peacekeeping troops, through police and gendarmerie, administrators, accounts, human rights monitors and aid workers' (Kaldor 2003: 156). 'The aim of such a service', she adds, 'is to protect civilians before, during and after conflicts' (2003: 156).

This vision for reconfiguring policing and military practice can be understood within a broader framework emphasizing an 'integrated human security' approach articulated by the Commission. In particular, the protection of people is seen to be accompanied by a future-oriented focus on protecting human rights, building peace and producing sustainable development in a variety of realms. These different yet complementary strategies of human security governance place an emphasis on harnessing local forms of knowledge and capacity within a cosmopolitan framework. That being said, there is currently a tension between the cosmopolitan vision and the state-centric vision of governance, expressed particularly in the war on terror, which privileges 'homeland' security and the deployment of military capacity that supports such security. Before exploring this issue, we discuss the protection of human rights, the building of peace and the development of societies as other modes of human security governance.

Protecting human rights

In addition to military intervention, there are several mechanisms aimed at ensuring domestic compliance with international human rights standards. Hampson and colleagues examine three: sanctioning, shaming and co-optation (Hampson *et al.* 2002). Sanctioning involves the denial of foreign goods and services to nations that are not providing sufficient protection of human rights. Shaming involves various educational and outreach activities aimed at exposing state practices that have violated human rights. Co-optation seeks to both establish international legal norms and establish or reform domestic judicial structures that are capable of enforcing such norms. While these authors emphasize the role of international agencies in the use of these mechanisms, they also mention the role of intergovernmental bodies and non-governmental bodies at the local and transnational levels. In Argentina, for example, various non-governmental and community-based groupings use a mix of sanctioning, shaming and other techniques that has played a role in profoundly shaping the human rights culture of the country. At the heart of this culture is the desire to maintain a collective memory of human rights violations in the past and present (see Cohen 1995).

An iconic expression of human rights-based governance from the 'bottom up' is found in Argentina's 'Madres de Plaza de Mayo' (Mothers of the Plaza de Mayo). This group was devoted to finding the bodies of their sons and daughters who had 'disappeared' during the military dictatorship. The report *Crime, Public Order and Human Rights* describes the impact of the Madres as they gained momentum in the late 1970s:

> Beginning during the military dictatorship, every week, dozens of women would gather in the Plaza de Mayo, in downtown Buenos Aires, carrying pictures of their loved ones who had been carted away by government agents. At first, police violently repressed these demonstrations, but over time, the political cost of such public, visible violence against women (and mothers) became too great. The women, whose maternal interest in the welfare of their disappeared children could not credibly be described as subversive, opened a path for others to criticise the abuses of the military regime. (ICHRP 2003: 54)

In Argentina generally there is a prominence of civil society groupings that are governing human security through their exposure and condemnation of state institutional violence. Another such example is the Centre for Social and Legal Studies (CELS), formed in 1979 to further advance the agenda of exposing particular instances as well as systemic practices of state institutional violence during Argentina's military dictatorship, and later on after the formal shift to democratic government. For CELS and other human rights groups it is important to prosecute those responsible for human rights violations during the dictatorship. This work continues today, along with efforts to prosecute members of police organizations and other state actors involved in 'post-transition institutional violence' such as police brutality, extra-judicial executions and excessive use of force (ICHRP 2003: 54–60).

As once described by an Argentine colleague processes of sanctioning and shaming perpetrators of human rights violations aim to 'structure their field of possible actions' for the future (phrasing that obviously draws on Foucault: see Foucault 1982: 221). In other words, shaming and sanctioning serve to induce future compliance with human rights norms by making it legally difficult, as well as publicly unacceptable, to undertake practices that deny such rights.

Another strand of human security governance within a rights-based approach can be found in 'Truth Commissions' in countries like South Africa and El Salvador (Hampson *et al.* 2002: 22). Truth Commissions are perhaps more accurately described as institutions for governing the future, that while concerned with establishing certain truths about past violations of human security are devoted to finding restitution and conciliatory solutions to such violations (Leman-Langlois and Shearing 2004).

In the case of South Africa, the Truth and Reconciliation Commission was established, in Tutu's words, as a 'third way of conditional amnesty', an alternative to, on the one hand, the purely sanction-based approach of Nuremburg and, on the other hand, the impunity-based approach that he described as 'national amnesty' (Tutu 1999: 33–4). For Tutu, it was important to remember and to establish the truth of what happened, and the perpetrators of past atrocities would be central to the process of establishing this truth and acknowledging the harm that was created.

According to Braithwaite (2002), the model of the Truth and Reconciliation Commission is both responsive and restorative. It is about '[s]peaking softly while being able to call upon a big stick' (2002: 196), or as Tutu describes it, 'the carrot of possible freedom in exchange for truth, and the stick was the prospect of lengthy prison sentences for those already in gaol, and the probability of arrest, prosecution and imprisonment for those still free' (Tutu 1999: 34). Within this model, victims of human rights violations have an opportunity to tell their stories and to participate in the development of reparative or rehabilitative plans of action (1999: 34).

Politicians agreed that amnesty applications would be considered as part of this process. This was not an uncontroversial consideration. While shaming was a component of the truth-telling process, in concert with Braithwaite's restorative theory, some may argue that shame isn't enough, and that survivors of human rights violations should have the opportunity to pursue civil and/or criminal damages. In response to this, Tutu explained that this 'third way' approach captures the African world view, *ubuntu*, which centres on the principle of forgiveness:

> *Ubuntu* is very difficult to render into a Western language. It speaks of the very essence of being human. When we want to give high praise to someone we say, '*Yu, u nobuntu*'; 'Hey, he or she has ubuntu.' This means they are generous, hospitable, friendly, caring and compassionate. They share what they have.

> It also means my humanity is caught up, is inextricably bound up, in theirs. We belong in a bundle of life. We say, 'a person is a person through other people'. It is not 'I think therefore I am'. It says rather: 'I am human because I belong.' I participate, I share. A person with *ubuntu* is open and available to others, affirming of others, does not feel threatened that others are able and good; for he or she has a proper self-assurance that comes from knowing that he or she belongs in a greater whole and is diminished when others are humiliated or diminished, when others are tortured or oppressed, or treated as if they were less than who they are.
>
> Harmony, friendliness, community are great goods. Social harmony is for us the *summum bonum* – the greatest good ... To forgive is not just to be altruistic. It is the best form of self-interest. What dehumanizes you, inexorably dehumanizes me ... *Ubuntu* means that in a real sense even the supporters of apartheid were victims of the vicious system which they implemented and which they supported so enthusiastically. Our humanity was intertwined. The humanity of the perpetrator of apartheid's atrocities was caught up and bound up in that of his victim whether he liked it or not. In the process of dehumanizing another, in inflicting untold harm and suffering, the perpetrator was inexorably being dehumanized as well. (Tutu 1999: 34–5)

The spirit of 'ubuntu' is captured in Braithwaite's proposal (Braithwaite 2006) for a restorative approach to peacebuilding, central to which is the development of nodal assemblages of state and non-state actors in processes of shaming and reintegration.

Building peace

The Commission on Human Security argues that states and the international community should be responsible for protecting people in conflict and for rebuilding societies. This cosmopolitan emphasis on providing security and protection, however, cannot, according to the Commission, be understood as a stopgap measure. Rather, '[t]he measure of an intervention's success is not a military victory – it is the quality of the peace that is left behind' (Commission on Human Security 2003: 57). The Commission challenges the established view that recovering from violent conflict and establishing peace is a linear process, beginning with stabilization, provision of high-priority humanitarian assistance, reconstruction and ultimately development. It argues rather that long-lasting peace can only be achieved through simultaneous processes of healing and institutional renewal that ultimately draw from the capacities and resources of local, national and global actors (2003: 57–8).

Processes of reconciliation (see Lederach 1997) and other measures for establishing peaceful communal relationships are therefore linked to community and social development. The Commission argues: 'Post-conflict situations provide opportunities to promote change, to fundamentally recast social, political and economic bases of power – opportunities for including the excluded, healing fragmentation and erasing inequalities' (Commission on Human Security 2003: 58). Part of this process involves establishing a 'new democratic political order' that is supported by the development of a strong state. Social capital must also be rebuilt as a means of healing communities and promoting trust. Furthermore, economic hardships in post-conflict situations must be averted, as '[e]quitable and inclusive economic growth is critical to promoting political and social stability, while enlarging opportunities for people' (2003: 58). In this vein, we return to Kaldor's argument that a new democratic political order must be based on cosmopolitan norms, institutions and practices:

> A politics of inclusion needs to be counterposed against the politics of exclusion; respect of international principles and legal norms needs to be counterposed against the criminality of the warlords. In short, what is needed is a new form of cosmopolitan political mobilization, which embraces both the so-called international community and local populations ... political mobilization ... has to override other considerations – geopolitical or short-term domestic concerns; it has to constitute the primary guide to policy and action which has not been the case up to now. (Kaldor 1999: 114)

As part of this vision, the legitimacy and authority of political institutions must be enhanced. Such institutions must operate with popular consent and must derive their authority from the rule of law (Kaldor 1999). The values of cosmopolitanism – central to which are 'tolerance, multiculturalism, civility and democracy' (1999: 116) – serve as the basis for what Kaldor describes as 'overriding universal principles which should guide political communities at various levels, including the global level' (1999: 116). It is indeed the case that some such principles are already captured in the current treaties and conventions of international law, and a 'cosmopolitan regime' would emphasize violations of humanitarian law as well as human rights law. While humanitarian law is concerned with violations during wartime, human rights law is concerned with abuses of power during peacetime as well, such as repression of citizens on the part of state institutions (1999: 116). While there have been 'tentative steps' towards a cosmopolitan regime, Kaldor argues that they 'conflict with many of the more traditional geo-political approaches adopted by the so-called international community which continue to emphasize the importance of state sovereignty as the basis of international relationships' (1999: 117).

In places where new wars are being carried out, Kaldor argues that it is essential to create the conditions for 'non-exclusive political constituencies' to develop and grow within a cosmopolitan normative framework (1999: 120). It is insufficient to undertake negotiations with warlords to the end of establishing some middle ground between 'incompatible forms of exclusivism' (1999: 121). While negotiating with warlords, 'mediators have to be very clear about international principles and standards and refuse compromises that violate those principles' (1999: 120). 'The point of the talks', she adds, 'is to control violence so that space can be created for the emergence or re-emergence of civil society' (1999: 120). Kaldor's dream for growing civil society is supported by evidence that in the midst of new wars, one can find the existence of 'islands of civility', consisting of people and places that espouse a cosmopolitan politics (1999: 120). These islands, she adds:

> … are rarely reported because they are not news. They involve local negotiations and conflict resolution between local factions, or they may involve pressure on the warring parties to keep out of the area …
>
> These groups represent a potential solution. To the extent that they are capable of mobilizing support, they weaken the power of the warring parties. To the extent that the areas they control can be extended, so the zones of war are diminished. They also represent a repository of knowledge and information about the local situation; they can advise and guide a cosmopolitan strategy. (Kaldor 1999: 121)

It is not enough, Kaldor adds, for the local knowledge contained in islands of civility to be valued and respected. These islands need to be protected and they need to operate in a climate of law and order. If states are not capable of providing this protection, then it is necessary for international organizations to intervene, whether in the form of sending in troops, funding reconstruction, or other support that is deemed necessary by local constituencies (1999: 122–4). Unfortunately, Kaldor writes, top-down diplomacy continues to dominate peacemaking efforts at the expense of deliberative processes that harness local knowledge and the contextual, tailored solutions that this knowledge informs (1999: 122–3). 'The point', she argues, 'is that local cosmopolitans can provide the best advice on what is the best approach; they need to be consulted and treated as partners' (1999: 124).

In his recent work on peacemaking networks (Braithwaite 2002, 2006), Braithwaite adopts Kaldor's cosmopolitan perspective in line with his work on responsive regulation and restorative justice. Consistent with Kaldor's emphasis on the protection and nurturing of 'islands of civility', Braithwaite (2006) argues that 'what is needed is a capability to escalate up an enforcement pyramid from non-intervention to dialogue and preventive diplomacy, to peacekeeping, to peace enforcement'. Also similar to Kaldor's articulation of 'cosmopolitan law enforcement', Braithwaite argues that the role of the police should be 'to ensure that islands of civility are not crushed' (2006). Police would be located within a responsive regulatory pyramid, willing and able to escalate up to more coercive interventions – such as mobilizing heavily armed peacekeepers – if restorative peacemaking fails (2006). 'The crucial role of the police', he adds, 'is not to create circles of reconciliation, but to secure the perimeters of such circles when they bubble up from civil society with support from outside NGOs, from the World Bank and others with the resources that count for peace' (2006). This is in essence Braithwaite's vision for 'networked peacemaking' which links and binds the knowledge and capacity of peacemaking nodes within an architecture of responsive regulation.

According to Braithwaite, traditional peacemaking approaches based on elite diplomacy do not provide for lasting peace (2002, 2006). Rather:

> Where there are injustices in terms of breaches of international law and injustices that are root causes of a war, the difference between a restorative justice philosophy and a philosophy of peacekeeping (as in simply ending the conflict) is that the restorative justice approach demands best efforts to right the wrongs, to heal the injustices. (Braithwaite 2006)

Restorative justice processes, he argues, should be established in islands of civility. As we discussed in Chapter 2, the restorative process involves stakeholders coming together to discuss the harm caused by an injustice and to explore ways of repairing those harms and addressing the needs of the victimized (Braithwaite 2006). Braithwaite's long-term vision is for islands of civility to expand by reaching out to zones of war through conversations that espouse the benefits of reconciliation:

> Mostly, perpetrators and their supporters will come from a different community than victims and their loved ones. So a conference convened in an island of civility, where some perpetrators live, would invite victims and their families from surrounding communities to hear the terrible truth uttered in hope of reconciliation. When that reconciliation happens, a bridge is built from the island of civility to the neighbouring community. If that neighbouring community's citizens believe they benefit from the ritual of healing, their perpetrators might be persuaded to offer up their truth, apology and gifts of repair to a third community. This is the restorative justice ideal: ripples of peacebuilding moving out from islands of civility. Local creativity, and familiarity with local custom, are crucial to turning ripples into waves of peace that wash across a nation. (Braithwaite 2006).

As with both Braithwaite's and Kaldor's analyses, it is not the case that local knowledge should be privileged over the knowledge possessed centrally or internationally by elite diplomats. Rather, one must 'defer to the local knowledge of the other' (Braithwaite 2006):

> The social engineers of statist diplomacy don't have enough local knowledge to understand the real conflicts that are touching people's lives. The conflicts on the ground are always more complex than their reifications, more rapidly changing than the intelligence reports from diplomats in air-conditioned offices can keep up with. Only indigenous ordering … to define the cross-cutting conflicts in local terms will deal with the local drivers of a war. Equally, there may be geo-political dimensions of the conflict that can only be understood in the language that is spoken in a meeting between major and minor state powers in the Office of the Secretary-General of the United Nations in New York. (Braithwaite 2006).

Developing communities and societies

As indicated earlier, an integrated human security framework calls for a coordinated set of processes for protecting and securing people, for providing humanitarian assistance, for building peace, and for reconstructing institutions and developing societies. This integrated approach is especially necessary in conflict-ridden societies, but it is applicable more generally to all parts of the world where human security is threatened. More broadly, the Commission emphasizes both elements of 'protection' and 'empowerment', which involve safeguarding people from those conditions that threaten their basic survival while creating opportunities in which human beings can grow and flourish (Commission on Human Security 2003: 3).

In understanding the development-based focus of human security governance, Duffield and Waddell (2006: 3–4) draw from Foucault's notion of a 'biopolitics of the human race' (2003) aimed at shaping the flow of events at the level of the population by acting on 'multiple social, economic and political factors that aggregate to establish the health and longevity of a population [which] appear at the level of the individual as chance, unpredictable and contingent events' (Duffield and Waddell 2006: 3). Accordingly '[s]ecurity… relates to improving the collective resilience of a given population against the contingent and uncertain nature of existence' (2006: 4). As part of this focus on improving resilience the Commission on Human Security argues for innovative forms of governance that manage market-related risks. According to the Commission:

> The central issue from a human security perspective is not whether to use markets. It is how to support the range of diverse institutions that ensure that markets enhance people's freedom and human security as effectively and equitably as possible – and that complement the market by providing core freedoms that the market cannot directly supply. (Commission on Human Security 2003: 75)

Macro-level initiatives focus on areas such as macroeconomic policy development, international monetary aid or foreign economic investment, while forms of local governance tend to focus on establishing non-market institutions and processes that provide for minimum standards of human security in times of crisis (alternative forms of insurance) while creating opportunities for people to build and exercise their capacities in fulfilment of their own personal development (2003: ch. 5).

Rankin notes that the emphasis on microeconomic development is indicative of a broader shift in thinking that favours market-based over state-based approaches to development (Rankin 2001: 18–19). 'Microcredit' programmes, for example, operate in different societies experiencing threats to human security due to developmental problems. Such programmes are designed to promote economic security by providing small loans to support micro-enterprises. A pioneering microcredit model comes from the Grameen Bank in Bangladesh, which gives small loans to support small-scale forms of self-employment. It focuses in particular on supporting women, based on the view that an improvement in their economic status will lead to improvements in their social and political statuses (Bernasek 2003: 369) and, presumably, their social and political security. The Grameen Bank model (among a wide range of models – see Yunus 2004) has been replicated in other places like Nepal (Rankin 2001).

Yunus (2004) indicates that while 'Grameencredit' initiatives vary in their local specificity, they converge on several core features. To begin with, credit is understood as a human right. The programme is targeted at those least able to access credit within established market economies, namely, the poor, with an overwhelming majority being poor women. The banking system itself is completely reconceptualized. Access to credit does not depend on the existence of collateral, and there are no contracts that are enforceable by law (Yunus 2004). While loans are mainly given for the purposes of supporting self-employment (2004), there is a range of other loans to support activities such as buying houses, building latrines and irrigating small gardens (Bernasek 2003: 373). The bank also links economic development to social development. For example, it provides a list of goals for borrowers and their families, such as committing to better sanitation, family planning, education and better nutrition (2003: 373; Yunus 2004). The bank also provides scholarships and student loans (Yunus 2004).

As of May 2006 the Grameen Bank reports having 6.39 million borrowers, 96 per cent of this figure being women. (Grameen Bank 2006). There are many other examples of microcredit/micro-financing models and institutional arrangements across the globe, such as the Self-Employed Women's Association (SEWA) in India (SEWA 2004; see also Datta 2003); the Working Women's Forum (WWF) in India (WWF 2004); 'cassies villageoisie' in Mali (see Cerven and Ghazanfar 1999); the Bangladesh Rural Advancement Committee's Income Generation for Vulnerable Group Development Program (IGVGD) (see Matin and Hulme 2003), Caja Los Andes in Bolivia (Vogelgesang 2003), the savings schemes of the South African Homeless People's Federation (South African Homeless People's Federation 2004); and the savings schemes of an alliance of three housing activist groups in Mumbai (the Society for Promotion of Area Resource Centres (SPARC), the National Slum Dwellers Foundation and Mahila Milan); in conjunction with other local organizations such as the South African Federation of the Urban Poor (FEDUP) and the global network of housing activists, Shack Dwellers International (SDI).

There is also an ambitious project stemming from the Microcredit Summit, held in February 1997, which consists of private sponsors (including Citicorp, Chase Manhattan and American Express), that pledged to reach 100 million of the world's 'poorest' by 2005 (Rankin 2001: 19; Daley-Harris 2003: 3). By 'poorest' is meant those that live in the bottom half of those living below national poverty lines, or those that constitute part of the 1.2 billion that live on less than $1 per day. This is done through credit for self-employment and other business services to the end of developing 'financially self-sufficient institutions' (Daley-Harris 2003: 3).

The microcredit and savings schemes movements are not without their critics. Rankin, drawing on her study of microcredit in Nepal, sees the movement as emblematic of 'prevailing neoliberal orthodoxy [that] has assumed a distinctively feminized character', whereby forms of grassroots governance simply express a 're-scaling of state power to the local level' (Rankin 2001: 19). She writes that as part of the neo-liberal project, women are being constituted as 'clients' who must be responsible for the economic security of themselves and their families, rather than being constituted as social citizens with rights and expectations *vis-à-vis* the state (2001: 19–20).

A different concern spans beyond the economic strand of development and relates more broadly to the 'security and development nexus' (Duffield and Waddell 2006: 18). It is a nexus that reveals deep tensions associated with an ongoing 'dialectic' (Kerr 2003) between human-centric and state-centric imaginings of security governance. As we see in the case of the war on terror, state government authorities have endeavoured to align, and render coherence to, seemingly incapable objectives: the protection of 'homeland' security largely through military might, and the promotion of human security through micro-level and macro-level initiatives for sustainable development (see Liotta 2002). While key policy documents like the 9/11 Commission Report (National Commission on the Terrorist Attacks upon the United States 2004: ch. 12) argue that such an alignment between state security and human development is possible and is indeed necessary as part of an integrated security governance package, there are others of the view that the war on terror is trumping the human security agenda (Duffield and Waddell 2006; Commission on Human Security 2003).

Duffield's work suggests, however, that even prior to the current war on terror, there was a shared agenda of Northern governments to promote what he describes as a global 'liberal peace', a political project aimed at developing and transforming societies – particularly those in the South – in the name of global security (Duffield 2001, 2005). As part of this agenda, 'security' and 'development' discourses have been merged. In particular, societal conflict came to be seen as causally linked to 'a developmental malaise of poverty, resource competition and weak or predatory institutions' (Duffield 2001: 15–16). In his book titled *Global Governance and the New Wars*, Duffield adds:

> The links between these wars and international crime and terrorism are also increasingly drawn. Not only have the politics of development been radicalised to address this situation but, importantly, it reflects a new security framework within which the modalities of underdevelopment have become dangerous. This framework is different from that of the Cold War when the threat of massive interstate conflict prevailed. The question of security has almost gone full circle: from being concerned with the biggest economies and war machines in the world to an interest in some of its smallest. (Duffield 2001: 16)

In the following section we briefly discuss the tensions associated with attempts to imagine global governance as a nexus between state security and human security.

The state security/human security nexus

In the face of the new wars, nations and localities understood as experiencing problems of underdevelopment, poverty and internal conflict are being constructed as potential threats to global security (Duffield 2001; Duffield and Waddell 2006). Subsequent to the terrorist attacks of 11 September 2001, this security-development nexus has intensified (Duffield and Waddell 2006). Development is not displaced as a modality of human security governance, but is rather conceptualized as a 'strategic tool in the war against terrorism' (2006: 11–12):

> The war on terrorism has had an acute impact upon human security as an evolving assemblage of global governance. The predominance of security concerns, especially homeland security, means that issues of global circulation – of people, weapons, networks, illicit commodities, money, information, and so on – emanating from, and flowing through, the world's conflict zones, now influence the consolidating biopolitical function of development. That is, security considerations increasingly direct developmental resources toward measures, regions and sub-populations deemed critical in relation to the dangers and uncertainties of global interdependence. (Duffield and Waddell 2006: 10–11)

This security-development nexus is focused on diminishing the conditions of possibility for dark nodes and networks of insecurity, including '[i]nsurgent populations, shadow economies and violent networks' (2006: 12), to flourish in weak and underdeveloped societies and to threaten global security as they cross national borders. One such condition of possibility is the capacity of violent leaders to attract poor and alienated populations who are experiencing injustice and exclusion in their homeland (2006: 12). According to a report prepared by the OECD Development Assistance Committee (DAC), terrorist leaders 'feed on' and exploit feelings of helplessness and alienation of marginalized groupings to the end of garnering support for their organizations (OECD DAC 2003: 11, cited in Duffield and Waddell 2006: 12).

Based on this view, development is concerned with 'offsetting alienation' through 'a complex set of biopolitical interventions' with the ultimate goal of building 'the capacity of communities to resist extreme religious and political ideologies based on violence' (OECD DAC 2003: 8). As part of these interventions:

> Education and job opportunities become key, reflecting the concern that the new global danger no longer necessarily lies with the abject poor, who are fixed in their misery: instead, it pulses from those mobile sub-populations capable of bridging and circulating between the dichotomies of North/South; modern/traditional; and national/international. (Duffield and Waddell 2006: 12)

The sustainable development agenda, which had previously centred on reduction of poverty for the world's poorest (2006: 13–14) was now aligned with a counter-terrorism agenda that targets 'transnational populations living in strategic regions' (2006: 13).

Given the fact that the war on terror now 'dominates national and international security debates' (Commission on Human Security 2003: 23), there are some who worry about the existence of what can be termed a 'new unilateralism', led by the United States in its capacity as the 'world's most powerful military actor' (Loader and Walker in press). From Kaldor's perspective, for example

> the global coalition constructed by the Bush Administration in the aftermath of September 11 was an alliance of states on the Cold War mode. It was not, despite the claims of some of the allies, a multilateralist alliance, based on international principles. As in the Cold War period, the criterion for membership in the alliance is support for America, not democracy or respect for human rights. (Kaldor 2003: 151–2)

Some aid agencies argue that the war on terror has compromised the human rights agenda. In particular, certain states have justified their lack of compliance with human rights treaties by appealing to the threat of terrorism. Not only is detention without trial being utilized in the US and Britain, 'many members of the global "coalition of the willing" have used existing legislation or passed new national security laws which, critics argue, have used terrorism as pretext for repressing legitimate internal opposition' (Duffield and Waddell 2006: 14).

It is also suggested that the traditionally apolitical character of aid work has been jeopardized by the war on terror. During the 1990s, military doctrine granted aid agencies a 'humanitarian space' (Feinstein International Famine Centre 2004, cited in Duffield and Waddell 2006: 14) within which the latter could conduct their activities while the former (the military) would only intervene as a last resort. It is argued that aid work has now been reconfigured in line with a broader political agenda of social reconstruction and democratization, as evidenced in the case of Afghanistan (Duffield and Waddell 2006: 14). While this 'draws [aid agencies] directly into a political process', the high level of insecurity in places like Afghanistan has served to draw the military into activities previously understood as humanitarian, such as the delivery of supplies and the repairing of infrastructure (2006: 14–15). 'The modern way of war', Duffield writes, 'not only requires the political and material support of "coalitions of the willing", containing and managing its inevitable humanitarian consequences has necessitated a growing interdependence between military establishments and the aid community' (Duffield 2002: 154).

The overall concern about the potential for the war on terror to undermine the progress made in human security governance is expressed vividly by the Commission:

> New multilateral strategies are required that focus on the shared responsibility to protect people. Considerable progress has been made in the 1990s – as exemplified by the prominence given to human rights and humanitarian action, as well as the efforts to deploy peacekeeping operations and rebuild conflict-torn countries. But the 'war on terrorism' has stalled that progress by focusing on short-term coercive responses rather than also addressing the underlying causes related to inequality, exclusion and marginalization, and oppression by states as well as people. (Commission on Human Security 2003: 24)

Of course there are many examples of tensions between human-centric and state-centric imaginings of security governance. The governance of illegal migration is one particular case. From a human security perspective, the global movement of people in the form of illegal migration poses clear threats to the health and income security of individual migrant workers, particularly those involved in domestic labour and in the sex trade. At the same time, illegal migration can be conceptualized as an attack on state sovereignty (Piper and Hemming 2004). This latter understanding supports initiatives to protect state borders that may trump concerns with the human rights of migrants themselves (Commission on Human Security 2003). 'In the name of preserving state security', the Commission writes, 'the detention of illegal migrants without due process is on the rise globally. People are frequently turned back by force at border points, returned to countries where their human rights may be at risk' (2003: 42). This is occurring in a world where 'the power to control borders is one of the few remaining prerogatives of the declining nation-state' (Weber and Bowling 2004: 197).

In relation to this 'expression of national sovereignty' (2003: 200), Bigo describes the formation of the European Union in terms of a 'fusion' of interests previously demarcated around nation-state borders (i.e. internal versus external security) (Bigo 2000: 82–4). Issues of domestic and foreign affairs have become reconceptualized within a broader discourse of European insecurity. In the process, domestic police agencies have been developing transnational networks, while military agencies focus inwardly 'in search of infiltrated enemies' (2000: 83). Bigo explains that there has been an 'intertwining of police and military technology (Haggerty and Ericson 1999); the armed forces have assumed more responsibility for the maintenance of surveillance of suspect populations (Sheptycki 1998a)' (Bigo 2000: 83).

It could be surmised that the governance of migration is taking place more in the name of state security than it is in the name of human security (see Angel-Ajani 2003). Further evidence of this can be seen in police organizations becoming more involved in the enforcement of immigration laws while other administrative agencies of the state have acquired coercive police-like powers in the same pursuit (Weber and Bowling 2004: 200). The constitution of 'transnational police networks and working cultures' (Loader 2002: 127) can be seen in what Weber and Bowling (2004) describe as 'migration policing', whereby police knowledge, and the imaginings of security they reveal, are extended outward to other institutions including state administration departments (e.g. immigration authorities) and transnational private policing agencies (Johnston 2000b). Weber and Bowling depict

> an ever-widening participation in [the] network of enforcement. Within Britain this already encompasses: private security firms involved in detaining and transporting asylum seekers and 'immigration offenders'; public and private employers required by law to check the immigration status of job applicants; haulage companies liable to criminal sanctions for transporting clandestine entrants across the English Channel; airlines subject to fines under the Carrier's Liability Act 1987 for allowing passengers to board flights bound for Britain with 'inadequate' documentation; and the population at large who are increasingly being encouraged to act as the eyes and ears of the police and Immigration Service. (Weber and Bowling 2004: 202–3)

The institutional networks of police cooperation are thus 'giving rise to new (hybrid) working practices and dispositions among officers moved by the common threat of the transnational criminal Other' (Loader 2002: 132).

Due to concerns with a new transnational 'criminology of the other' (Garland 1996), it is unsurprising that some observers are cautious about the degree to which securitization should take place. In fact, Buzan and colleagues argue that 'security should be seen as negative, as a failure to deal with issues as normal politics. Ideally, politics should be able to unfold according to routine procedures without this extraordinary elevation of specific 'threats' to a prepolitical immediacy' (Buzan *et al.* 1988: 29). Kitchen (2001) agrees that 'desecuritization is the optimal long-range option' (Buzan *et al.* 1988: 29):

> if enough people say something is a security issue, it becomes one. Once something has been designated as a security issue, we have given ourselves the liberty to treat it with extraordinary means. In old-style security thinking, that can mean constructing an enemy and using military metaphors to mobilize the population against the threat. If your country is being invaded by hostile troops, this may be the correct response; if your citizens are starving due to the erosion of arable land, it is probably not. (Kitchen 2001: 3; see also Møller 2000: 43)

In contrast to those who seek to widen the security agenda, Kitchen therefore argues that 'human well-being should replace the concept of security in all instances other than threats to states' (2001: 4; see also Wæver 1995: 58).

In short, it could be said that the dream of human security governance is incomplete in its realization, as are the dreams of state-centric and community-centric imaginings. Rather, such dreams coexist, in some cases in very uneasy ways. Such dreams may also be re-imagined in ways that their current visionaries have not predicted or planned for. Scholars like Held (2003, 2004) and Kaldor (2003) nonetheless remain hopeful that a cosmopolitan ethic is the most sensible and desirable normative driver of global governance. As a slight alternative, Loader and Walker have begun to develop the notion of an 'anchored pluralism' (Loader and Walker 2006, in press) that endeavours to re-imagine the state as a 'necessary virtue' (2006), and in particular as a 'constitutive political community which is involved not only as a last resort of coercive authority, but also – since the two activities are inextricable – *both* in instrumental ordering work *and* in the work of cultural production of social identity' (2006: 193). Loader and Walker add that while states must be 'the primary motors of common action and sources of institutional initiative' there must be a 'principled recognition that there are two levels at which we can think of security as a thicker public good which are not reducible to one another but which need different institutional forms for their articulation' (in press). In other words, a 'thick' conception of the public good must be expressed at the state level as well as at the global level, resulting in a 'pluralism of levels of the public good' (Loader and Walker in press).

Suffice it to say that there are competing normative agendas for the future of global governance, each of which expresses different imaginings of states and their roles within a transnational architecture. What we do know is that while 'states remain important … they have been drawn into multi-level and increasingly non-territorial decision-making networks that bring together governments, international agencies, non-governmental organisations, and so on, in new and complex ways' (Duffield 2001: 11). We will address issues pertaining to the global regulation of security governance within a nodal world in Chapter 5.

Conclusion

If we return to our Latourian understanding of how power is exercised, one could understand the dialectic between state-centric and human-centric imaginings in terms of a kind of power play between different actors involved not only in the speech act of securitization, but also in the enrolment of other actors in a particular dream of governance. In this way, one could come away from the above discussion of human security and global governance with the observation that some actors are simply 'weaker' than others in their attempts to shape security governance towards instrumentally and normatively desirable ends. This point was made by Braithwaite and Drahos (2000) in their analysis of weak and strong actors in the field of global business regulation.

In the next chapter we take this point forward and shift to an explicitly normative discussion of how a nodal governance perspective might inform efforts to enhance the power of weak actors, including those from 'islands of civility', in reshaping the security field towards possibly new and different ends, both in instrumental and normative terms. This perspective explores ways of identifying and harnessing nodes – as sources of knowledge and capacity – that are often excluded as decision makers and practical actors in the production of diverse security goods.

Chapter 4

Responding to governance deficits

> The ideas, the networked connections, the new directions, must come from below ... A dense network of decentralized governance is impossible for anyone to understand synoptically. We only understand bits of the network that we can monitor directly. While governance cannot encompass synoptic planning, actors can govern nodally. They can see enough of what is going on to clarify that if they can get other key actors who control different strands of a network of governance to collaborate at a particular node, those strands can be tied together into a web of controls. (Braithwaite 2004: 308)

Introduction

The preceding chapters have sought to illustrate the nodal character of security governance. While state nodes remain an important part of this fluid nodal architecture, non-state nodes, including business corporations and non-governmental organizations, now play a major role in imagining, directing and providing security goods. Security governance, depending on one's place within nodal relationships, produces both good and bad consequences. For state nodes, the capacity to enlist others in the pursuit of state-centred objectives serves to enhance their resource base and their strategic capacity. For corporate nodes, the capacity to enlist state and other nodes in realizing their corporately defined objectives achieves high levels of self-direction and autonomy. At the same time, for those with limited capacity to enlist others – whom Braithwaite and Drahos characterize as 'weak actors' (Braithwaite and Drahos 2000; Braithwaite 2004) – nodal governance, as presently configured, leads to less rather than more security.

Put simply, some people have done better out of nodal governance than others, just as some people did better out of other forms of governance. Not surprisingly the fault lines are closely associated with wealth. The rich have done better than the poor. Of course, it is not simply access to economic capital that allows nodes to shape the security field. In this vein, Dupont (2006a), drawing on Bourdieu (1986), describes the field of security in terms of nodes whose position in the field is shaped by the amounts of various forms of capital (symbolic, cultural, political, social and economic) they possess (see also Chan 1996, 2001a; Chan *et al.* 2003). From this perspective, the power of nodes – and the legitimacy of the claims they make to their audience – depends on the types and amounts of capital such actors possess or are capable of amassing. In a nodal world, deficits in the capacity to shape the field of security have had a differential impact on individuals, collectivities and states throughout the world. This is true of the governance of security and it is also true of governance generally.

The central normative question we address in this chapter arises directly from this observation. Our question centres on how nodal relations could be transformed to improve governance processes and outcomes for weak actors. In answering this question we take for granted that nodal governance will be with us for some time, though its operations will evolve and change across time and space. Given this, we should seek nodal solutions to these nodal problems. Our reasons for working within a nodal framework, as we have noted, do not simply have to do with the likely persistence of this form of governance as we conceive it. We believe that our nodal reality carries within it new opportunities for finding solutions to old and enduring problems. Nodal governance provides important opportunities that we should identify and harness. There is a considerable body of work that has sought to do just this – through state and non-state nodes, below states, within states and across states (McGinnis 1999b).

This normative approach is a feature of the work of a group of scholars whose empirical interests range across a gamut of terrains of governance including development, health, security, business, environment, knowledge and intellectual property associated with the Regulatory Institutions Network (RegNet) at the Australian National University. A significant theme of RegNet's research, policy and practical interventions has been, and continues to be, the exploration of possibilities for realizing established democratic values within nodal or decentred governance contexts at macro and micro levels. A foundational claim of this body of work is that normative opportunities for realizing democratic values can be identified and crafted within a nodally governed world.

The approaches being explored through RegNet often run against the flow of mainstream thinking; they do not run along neat Hobbesian tracks that assume that what is required normatively is simply more effective and accountable Leviathans. That is, they do not assume *a priori* that the best or exclusive way to respond to the problems we have identified is to create better Leviathans either at state or supra-state levels – although this solution is not excluded. Their thinking has been more inclusive. These scholars accept that more effective and accountable state and supra-state governments can and must be, under the right circumstances, a good thing. But they also accept that there may well be other, and perhaps better, ways of creating both effective and democratic governance. In taking this stance they look critically but with an open mind at a whole range of possibilities. In particular they adopt Rose's (1996) view that a 'simple dismissal' of complex forms of governance, that have had mixed consequences, is likely to be counterproductive.

In building on this body of work, this chapter focuses in particular upon, and seeks to develop, the work Braithwaite and Drahos have undertaken in exploring global business regulation (Braithwaite and Drahos 2000). In particular, we focus on a normative extension of Braithwaite's (2004) articulation of a series of 'methods of power' that he conceives of as 'weapons for the weak'.

In developing our arguments we take the position that while nodal governance has undoubtedly created a variety of problems and exacerbated well-established ones, it is not its nodal features *per se* that are the problem but the particular way in which nodal governance has evolved. Furthermore, as we have just suggested, given the history of centralized forms of governance that seek to rein in pluralism we are less optimistic than many mainstream thinkers about the possibility of creating centralized forms of governance that do not reproduce, and indeed do not exacerbate, the problems that concern us here. We believe that collectivities are more likely to successfully address governance deficits if they concentrate on reshaping nodal governance rather than seek to eliminate its nodal features. This requires both understanding how power is exercised within nodal governance as well as locating and exploiting opportunities this analysis identifies for responding to these deficits.

The overwhelming story of our nodal reality is that ultimately it is nodal power that talks. Individuals, collectivities and formal institutions access nodes that are effective in shaping flows of events that are important to them and that they can directly influence. States themselves constitute a nodal assemblage that can and does do this. This has been recognized in conventional thinking. This thinking has focused attention primarily on states and the assemblages of nodes that make up states within a hierarchical structure of governance. This picture has become more complicated as other nodes have become more active as governing entities. Sometimes these other nodes have been enrolled and recruited by state nodes as part of rule-at-a-distance arrangements. This certainly happens but it is not all that happens. There are other governing hierarchies that engage other nodes, including state nodes, as part of their own rule-at-a-distance strategies. An example is the way in which corporate governance operates to pursue its objectives (Williams 2005a). This mix of overlapping hierarchies of governance gives nodal governance its complex character of up and down and across arrangements.

In addressing deficits in the capacity to shape the nodal field of security governance it is essential, as Braithwaite notes at the head of this chapter, that 'the ideas, the networked connections, the new directions, must come from below'. It is ordinary people, both individually and within collectivities that bear the burdens of weak governance who must explore how to gain a greater and more effective role in its operation.

A conventional solution to finding ways of enabling bottom-up solutions would be to try to ensure that established democratic institutions work more effectively in allowing excluded communities to have a greater say in governance. While we certainly support such approaches we argue here that it is necessary to move beyond this by searching out and deploying nodal possibilities that exist outside of established democratic institutions. In our discussion below our focus is on what the weak can do to catch up with what the strong have already done and do well, namely to gain and then maintain access to nodes, including but not limited to states and their agencies.

Practical strategies that enable weak actors to gain better access to governing nodes are being explored in a variety of contexts through what Dorf and Sabel (1998) would depict as 'democratic experimentalism'. This Deweyan trial and error experimentalism (Karkkainen 2004) works from the bottom up and is the approach that we have pursued with others within communities of weak actors in Argentina, Australia, Canada, South Africa and more recently in Brazil (for a review of this approach see Johnston and Shearing 2003). The question that has motivated us is precisely the one that has motivated much of the work of Braithwaite and Drahos. Their body of work provides an important comparative resource that we draw upon in this chapter.

Braithwaite (2004) has sought to synthesize the lessons of this work by articulating a series of 'methods of power for development' to be used as guides by weak actors as they seek to take better advantage of the nodal opportunities available to them in shaping the flow of events. In developing his arguments Braithwaite draws upon understandings of the ways in which strong players have secured and maintained their positions of strength. In using the experiences of the strong to provide insights for the weak, Braithwaite has adopted a tack very similar to the one we have used in the research we have undertaken. We seek in the following pages to combine our experiences with his to articulate strategies for responding to governance deficits whilst promoting core democratic values. We conceive of these as methods of power for re-imagining the governance of security.

Methods of power

Concentrate power nodally and use it to steer governance

Braithwaite's first method of power is: 'Concentrate nodally the power under your direct control' (Braithwaite 2004: 311). We all have some power to shape the flow of events and much of this is under our direct control. So it is certainly possible to concentrate this power. Just what the power under our direct control is varies considerably. Braithwaite points out that one of the characteristics of weak actors is that they often do not have much conventional power to concentrate at organizational sites of governance. For example, developing countries do not have much economic power, nor do they have much military power.

Developing countries do, of course, have some economic power and some military power. They also have other power assets that are often unrecognized because they are so focused on the sorts of power resources that strong states deploy. They can and should concentrate whatever sources of power they have nodally in ways that will strengthen their bargaining position. This is certainly worth doing no matter how limited their power resources. For example, as Braithwaite notes, developing states should do as much as they can to strengthen their capacity as organizations so that their agencies become more effective nodes through which to concentrate what state power they have. Also, as Braithwaite points out, it is important to create nodal assemblages with other developing states that enable them to coordinate their actions. In making this suggestion, he notes how successful state coalitions have proved both for strong and for weak states.

But states are, of course, not the only nodes where power can be concentrated, nor is the power to be concentrated only within actors' direct control. There is much more that can be nodally concentrated and many more nodes to concentrate it through. Power can also be, and indeed normally is, exercised through enrolling others, as the neo-liberal mentality of governance has emphasized. For Braithwaite enrolment is a key feature of 'networked governance'. Network governance, as Braithwaite conceives of it, is less about the flows of information that Castells (2000) has drawn attention to, and much more about the enrolment of others to form action networks, or what following Foucault's usage we might think of as 'power assemblages' (for a discussion see Rose 1999: 190 and O'Malley 2004: 24). Drahos calls the nodes through which enrolled power is concentrated 'supra-structural nodes' (Drahos 2005a: 419). An example Drahos provides is the policy advisory committees made up largely of business interests that, through their peak bodies, advise and direct the US government on trade negotiations. These nodes concentrate enrolled power and then seek to use the US government as a powerful node through which directions are then transmitted to the agencies that wield state power (2005a: 419).

In our own work (Shearing and Wood 2000, 2003a, 2003b; Wood 2004; Burris *et al.* 2005; Shearing and Johnston 2005; Burris 2006; Wood and Font in press) we have witnessed how people within poor communities have created nodes through which they can concentrate the power of their knowledge and resources to govern their security. In South Africa they are called Peace Committees, and they are now placed in a position where the public police see them as an asset that they can enrol. However, as Peace Committees have established themselves as nodes that steer rather than simply row, they have been able to use their power as a bargaining chip in negotiating with the police. This has led to a process of mutual enrolment. As the police have sought to enrol the Peace Committees, the latter have, in turn, used their power to enrol the police in furtherance of their governance objectives. The superstructural nodes that have been developed to do this are Community Peace Centres, which operate as nodes of mutual enrolment. In promoting their objectives this has been done through a superstructural node in the form of the Community Peace Programme (CPP). The CPP, which has been a part of the University of the Western Cape, has used the kudos of the university to bargain on behalf of Peace Committees with the police. This bargaining has enabled a process of mutual enrolment, a 'partnership' where each partner brings power to the table. As a consequence, the relationship established is one where two parties are contesting amicably for control of governing agendas.

The Community Peace Centre project is not simply a neo-liberal partnership where the police are 'responsibilizing' community members to do their bidding. Fung has described similar processes related to the governance of security and education as examples of a 'democratic bargaining' within the US (Fung 2001, 2004). Similar examples can be found in a whole variety of environmental arenas under the signs of 'collaborative management' and 'collaborative governance' (Selin and Chavez 1995; for a discussion of this literature see Karkkainen 2004). What is clear in all of these examples is that actors can and do access power and concentrate it through nodes. They then use these nodes to establish themselves as players who steer governance. They allow themselves to be enrolled but trade this enrolment for enrolment gains of their own. In doing so they use the enrolment of others to ratchet up their own power to become stronger players. This 'game' of enrolment is constant and gains can easily be lost if they are not consolidated and ratcheted up further – this ability to consolidate, Braithwaite and Drahos (2000) argue, is an important feature of strong players. Drahos (2005b) has shown how in the case of trade, developing states that fail to build on the gains they made soon lose their advantage.

The creation of Peace Committees as nodes has enabled very weak actors (in this case people living in 'shacks' who are thought of as the poorest of the poor) to turn their resources (particularly time and knowledge) into power and to use this to gain more power by enrolling others. In doing so not only have they recognized that they do indeed 'concentrate nodally the power under their direct control', but they can participate in concentrating and directing the power of others through enrolment by participating in power assemblages. This has enabled them to ratchet up their power. We have already mentioned similar examples of weak actors creating and accessing nodes in the activities of savings and loan schemes within very poor collectivities such as the work of the Grameen Bank and Shack Dwellers International and its affiliates. The way in which the Fair Trade movement and the suppliers that use it have moved to gain and use nodal power to strengthen their trade position provides another example.

In comparing our work on 'microgovernance' (Burris 2004) with the analysis of Braithwaite and Drahos, we note the shared observation that every set of actors has context-specific opportunities to engage in nodal governance strategies. Strong states have opportunities that weak states do not. Similarly, weak states have opportunities that strong states do not. Private corporations have opportunities that other actors within civil society organizations do not and vice versa.

One of the enrolment opportunities available for poor players operating at the level of community is their potential access to non-governmental organizations (NGOs) as enrolees. To the extent that they are able to enrol NGOs they can often significantly ratchet up their resource base and their access to expertise that they can use to build upon and complement their local knowledge. By concentrating the power resources they have available in nodal assemblages and superstructural nodes weak actors can also enhance their ability to shape the flow of events and punch above their weight. What is important about the method of concentrating power nodally is not simply what they might accomplish directly but how incremental gains can reconfigure the landscape of governance so that new possibilities for new resources and new enrolments are created.

The game of power is very often, as Braithwaite and Drahos make clear, a game of pulling oneself up by one's bootstraps. Bootstrapping is not simply a method that weak actors can and should use but has proved to be a fundamental feature of the strategies of the strong – indeed the strong are very often as strong as they are precisely because they have bootstrapped successfully. For example, the nodal power that international companies (who now own what was common property for millennia) now exercise was accomplished through countless small initiatives. In retrospect, what was accomplished seems straightforward. Prospectively, accomplishing 'ownership' of natural genetic sequences must have seemed a daunting goal. This required not only creating a particular form of ownership, namely the ownership of knowledge that was for much of human history thought of as a common asset of all of humankind, but also the achievement of this ownership through complicated connected sets of intellectual property laws and enforcement arrangements. Retrospectively, how this was accomplished may appear obvious, and perhaps even inevitable. Prospectively it would not only have seemed daunting but would have been very difficult to even imagine as a possibility (Braithwaite and Drahos 2000; Drahos and Braithwaite 2002).

Weak actors, despite the myriad ways in which they are constrained and disadvantaged, still have enormous opportunities available to them for building upon, accumulating and exercising power. This requires imagination, hope, resilience, determination and persistence (Cartwright 2004). These attributes can be used not only in concentrating the power of nodes, but also in shifting the ground they operate on so as to make the resources directly under their control more useful as power tools. This requires institutions that bring together sets of actors through processes of 'deliberative polyarchy' (Dorf and Sabel 1998). While weak actors have enormous understanding of how this can be accomplished, in some fields they are often considerably less accomplished at doing this on the terrains where the strong operate. The weak are thus frequently out-manoeuvred. This is a critical area in which weak actors can enrol the support of others, in the same way as strong actors have done, to assist them in reshaping the landscape of governance.

Recognize and use all your power resources

The second power method is related to, and broadens, a method identified by Braithwaite: 'Have a big stick and threaten to use it' (2004: 320). Or expressed differently: 'Speak softly but hold a big stick'. This idea of escalating sanctions that Ayres and Braithwaite (1992) speak of as a 'regulatory pyramid', involving the 'responsive' escalation of force from soft methods used initially to harder and harder ones in response to continuing non-compliance, is a very old and well-established strategy. In showing that one has a big stick one announces the possibility of escalation and invites others to take this into account in making decisions about their behaviour. What one hopes to accomplish by doing this is to persuade them to respond positively in the face of softer methods rather than wait until heavier methods, which might be more damaging to them, are applied. As Braithwaite and Drahos (2000) note, strong actors are often strong not because they are always hitting others with big sticks, but because they have big sticks (and often bigger sticks than others) that they wave around and are willing to use. As Braithwaite (2004) points out, weak actors also have access to big sticks although they might not always know this. In making this point, he cites Monbiot (2003) who points out that the high levels of debt possessed by some developing states constitute a big stick which they could use more effectively in the governance game.

It is perhaps too limiting to think in terms of escalating from small to big sticks even though it does describe what is and can be done. In understanding how one can move beyond this way of thinking it is helpful to examine what has been happening in many police organizations as they have sought to reconceptualize their access to resources and the ways in which they can use them to promote compliance. For many decades, police organizations thought about how they should use their resources in terms of a regulatory pyramid. They conceptualized this in terms of a 'continuum of force'. This idea encourages police to speak softly, to carry a big stick (in the form of a firearm that can be used to kill others) and to escalate up from soft talk, through physical constraint, through to deadly force in ways that respond to the levels of compliance to their directions.

Police forces around the world have recently begun to rethink the ways in which they can and should use the power resources available to them. They are now conceptualizing their decision-making in more horizontal terms. For example, the Canadian National Use of Force Framework, which has been operational since 2000, conceives of a police officer not as escalating up a pyramid of power resources – starting at the bottom and moving upwards until compliance is achieved – but as an actor with a gamut of resources at their disposal that should be accessed in response to the nature of the situation facing them: 'The force options may be used alone or in combination to enable the officer to control the situation … The dynamic nature of the situation requires continual assessment, therefore, the force options selected may change at any point' (Canadian Association of Chiefs of Police 2000: 10).

Within this conception the police officer is encouraged to be responsive, but to be so in ways that recognize that what works will depend on the situation. That is, what is 'hard' and what is 'soft' (that is, the most persuasive power resource) will vary from situation to situation (see 2000: 11). Thus, while waving around and using big sticks like guns in some situations might be very effective, in other situations they might be much less effective than other seemingly 'softer' power resources. Sometimes 'softer' resources should be used first, but in other situations it might be better to use 'harder' resources first and then shift to 'softer' ones if this is likely to be more effective.

Within this decision model, power resources are not depicted as hierarchically organized but rather as available on a horizontal plane. One can start with any of these resources in creating compliance and move through them in any combination and in any order. The officer is required to assess what is best, and normatively most appropriate, for a particular situation. This includes, but is not limited to, ordering resources in a pyramidal manner. The intention behind this model (and the reconceptualization it embodies) is to expand the governance imagination of police officers, who are encouraged to operate with greater fluidity and agility across a wide range of resources.

There is much that weak actors can learn from this. Through watching and submitting to strong actors they can easily limit their imagination as to what resources are at their disposal. Because they do not have the big sticks that the strong have, and use, they can be persuaded (and persuade themselves) that they do not have at their disposal effective power resources. By taking a leaf out of the book of contemporary police organizations, and realizing that what is and is not an effective power resource varies considerably from situation to situation, they can expand their imagination considerably. When this happens they will discover, as Monbiot suggests, that they have many potentially powerful resources available to them that can greatly enhance their strength.

In our work with very poor communities we have found that once people get together within deliberative forums in which they are encouraged to stretch their imagination, they discover that not only do they have many resources available to them that can be used to govern effectively, but they have resources that can be used to enrol more powerful actors such as the police and align the decisions of these powerful actors with their agendas. For example, South African Peace Committees have discovered that they can help alleviate the budgetary crises that constantly beset police organizations by taking a host of problems off the police agenda and dealing with these problems themselves. In doing so they take a leaf out of the book of corporate actors who have long realized that they can use private security to gain greater autonomy while at the same time persuading police organizations to support them in doing this (Shearing 1992; Johnston 1992; Williams 2005b).

While weak actors should certainly be aware of and learn from the strategies used by strong actors, they should not allow their imagination to be limited by these strategic options. In developing their strategies it is important for weak actors to pay attention to and cultivate their own special strengths. There is no doubt that strong actors are strong because of their access to particular sets of resources that have worked well for them. But this should not detract from the fact that the weak have many resources that can be mobilized to create strength. We are reminded here of Granovetter's (1973) argument about the strength of weak ties within networks. Actors in weak positions often have access to many weak resources, both directly and through enrolment, that can be brought together to provide considerable strength. One of these resources is strength in numbers, provided these numbers can be made to count.

This second method has drawn upon Braithwaite's third proposition: 'Have a responsive regulatory strategy' (2004: 322). It suggests that it is just as important to be responsive to one's own situation and capacities as it is to be responsive to the reactions of others in deciding how best to exert one's power.

Focus on nodes where one can be creative and assertive

Braithwaite argues that shifting governance in new directions requires great energy and imagination. This leads him to state another method of power: 'Be creative and assertive at nodes of networked governance'. When one is participating in nodes where others are active, it is vital to be creative and assertive in realizing one's own agendas. In developing methods of power for the weak we suggest that this method needs to be broadened somewhat in ways that recognize the structuring features of nodes. If weak actors engage in governance in and through organizational nodes that others, particularly rich and powerful others, have established and work through, they will typically find themselves at a disadvantage because they will find that the agendas of others have been built into nodal processes. As a result, simply participating in a node can in fact work to reduce power because the agendas of others are foisted upon one. Perhaps the clearest example of this is the participation of the colonized in the institutions of governance of the colonizers, even when those institutions claim to recognize and to support indigenous institutions of governance.

Actors who wish to resist this can, and often do, establish nodes to work through that fit with their purposes. Examples can be drawn from the history of private security where companies took charge of their own security, not because they were unable to persuade state agencies to become involved, but because they wished to take charge of matters and not work through existing institutional arrangements that were structured to promote agendas that differed from theirs

(Johnston 1992). This, as the case of private policing makes clear, is not a matter of legitimate versus illegitimate objectives being pursued, as both sets of objectives may be perfectly legal, but rather whose governing agendas should take precedence (Shearing and Wood 2003a).

This issue of who steers governance is vital to weak players. A method for resisting the agendas of others is to create nodes, and to focus on acting through nodes, where processes are favourable to one's agendas. Again the experience of Peace Committees makes very clear that creating and participating within nodes where one can be assertive and creative is likely to prove more helpful than struggling to be assertive and creative in nodes where the tide is always flowing against one.

In the context of global business regulation Braithwaite and Drahos (2000) pay considerable attention to the ways in which the United States, and those who enrol its agencies, shift nodal forums (see also Drahos 2005a) when they find the tide is flowing against them. As we have noted, this is precisely what corporations have done by hiring their own police in the form of private security. It is also precisely what Peace Committees have done in establishing their own conflict resolution mechanisms and then inviting the police to cooperate with them. Indeed, Braithwaite identifies the capacity to 'shift forums' as another method of power (2004: 327).

Concentrate knowledge at nodes

The issue of what knowledge is required for governance, where it is located and who has access to it, is critical in understanding governance processes. In calling for a method that 'concentrates technical competence in nodes' (2004: 326), Braithwaite identifies not only the crucial place of knowledge, but also access to knowledge. He argues that knowledge is often organized through expertise and that this expertise requires an organizational site through which to operate. The use of expertise in the form of technical competence is clearly vital as a source of strength and one that can enable weak actors to become stronger players mobilizing expertise at nodal points. Clearly, expert knowledge and the ability to secure and to focus it is vital if weak actors are to strengthen their governance ability. However, once again we suggest that it is useful to broaden Braithwaite's conception.

The usefulness of broadening this method is most clearly apparent when the weak actors involved are local groups whose particular power advantage is often their access to local knowledge or what Nygren (1999) terms 'situated knowledge'. Other actors, who may well be strong actors, typically seek to gain access to situated knowledge through enrolment so that their agendas can be realized. An example from the field of policing might be the use of local community members to provide intelligence to police organizations. The encouragement of victims to call the police to report their victimization (9-1-1 in North America), as well as programmes such as 'Crime Stoppers', provide examples within the arena of state policing. Similar programmes exist within non-state policing, particularly with regard to theft within large business organizations. Another apt example is the encouragement of whistleblowers within securities markets.

Weak actors can, of course, use their situated knowledge to promote their own agendas. They can use this as a bargaining chip in negotiations with strong actors or they can do so by creating nodes whose agendas they control where their knowledge is valued. Again the work of Peace Committees is instructive. Communities in South Africa are using Peace Committees as nodes where their knowledge counts as places to pursue local agendas. This knowledge is used to resolve conflicts in ways that local people find acceptable. In addition, Peace Committee processes are used as nodes where agendas for local development projects can be established and implemented. The fact that this knowledge is concentrated through nodes enables these actors to enrol others, in particular governments and government agencies as well as international donors, to realize their agendas.

These local actors also use Peace Committee processes as a nodal focus through which to draw in the situated knowledge of others to assist them in realizing their agendas. The knowledge contained in nodes is therefore not simply a resource that stronger actors enrol to pursue their agendas, but a resource that is used to set governance agendas for weaker actors.

Locate resources at nodes

As we have suggested throughout this book, one of the key developments within the governance of security has been the location of resources in nodal locations outside of state governments. This is a central strategy that corporations and non-governmental organizations in particular have used to create and operate nodes of governance within and beyond states. These nodes collect and pool resources (various forms of capital) that allow them to pursue agendas that may or may not be aligned with state agendas. A simple example of the way this is done is the employment of private security to govern mass private property in ways that promote corporate agendas (Kempa *et al.* 2004). A more complex example can be found in the nodal assemblages of NGOs created to advance human rights and reduce state institutional violence.

Strong actors, however, are very adept at 'jockeying for position' (Dupont 2006a) in an increasingly nodal field of security governance. If weaker actors become effective in amassing resources/capital, stronger actors respond through attempts to consolidate and ratchet up their own resource concentration. In his study of police leaders' 'power struggles' in the field of security, Dupont writes:

> The governance of security is not only influenced by trends such as the pluralization of actors or the commodification of service provision, but also by power struggles that result from those trends. As the number and capacities of alternative security providers expand, police organizations are inclined to contest the legitimacy of resources deployed by the newcomers and adjust their own capital allocation strategies accordingly in order to maintain their status and position at the centre of the field. This does not imply that they seek total control of the field and that new entrants cannot carve a niche for themselves or even appropriate a significant share of activities – in fact, police organizations also purchase services from private security providers ... But this seems possible only so long as those new entrants limit their ambitions or do not openly challenge the dominant role of the police. Meanwhile, other actors such as governments, the media, unions and the public will intervene in those struggles, and use their own specific resources to further their interests. (Dupont 2006a: 104)

It is important for weaker actors to amass different forms of capital (i.e. symbolic, cultural, political, social and economic) and to jockey for position in ways that recognize the jockeying done by others. Again Peace Committees provide an example of the way in which weak actors can amass and locate resources at nodes. In financial terms, the resource pool that is concentrated at nodes has both an in-kind and a financial component. The in-kind contribution is the time, capacity and knowledge that weak actors concentrate through these nodes. The financial 'tax' component comes from block grants provided both by governments and by international donors. In other terms, the very functioning of the Peace Committees serves to consolidate and leverage social, economic, symbolic and possibly other forms of capital that, as Braithwaite might express it (2006), would be otherwise invisible to community development experts from the vantage point of their offices.

Promote deliberative processes within nodes

If the knowledge concentrated at nodes is to be useful for auspices and providers working through those nodes, this knowledge has to be considered, conclusions have to be reached and then acted upon. If this is to take place self-consciously in ways that enable these actors to act on this knowledge effectively, it is essential that deliberative processes exist that enable this knowledge to be considered and challenged. There is a very sizeable literature on this that there is no need to canvass here, ranging from considerations of the ideal institutional arrangements to promote deliberative processes to accounts of particular processes and the ways in which they work (Dryzek 2000).

An essential feature of the Peace Committee model has been the use of deliberative processes that enable local knowledges to be articulated, considered and related to expert knowledges. These processes encourage the development of a consensus with respect to what course of action should be taken to shape the governance of security and local development more generally.

Democracy in nodal governance

To the extent that weak governmental actors are able to increase their control over the steering and the provision of governance through nodal access, the ideals of democratic governance are likely to be enhanced. Widespread nodal access operates to introduce more voices from more constituencies into the public sphere. Voices alone, however, will not ensure that decisions taken will represent all of these voices. Established institutions of democracy provide ways of coming to decisions in the face of conflicts. The ideal that lies behind these methods is that the voices that constitute a majority within some polity (subject to constitutional constraints that express what are thought of as widely held transgenerational values) will be the ones that carry the day.

With very few, if any, exceptions, powerful interests have been able to 'game' these institutions and processes in ways that enable their voices to carry considerable weight no matter how small their populations may be and no matter how badly their proposals might fare within ideally constituted deliberative forums. Much thought has been and continues to be given as to how these game-playing strategies might be curtailed. These have led to a constant stream of proposals for institutional reinvention and reform initiatives that arise from these.

Following Braithwaite and Drahos we have been arguing for a set of strategies for levelling the democratic playing field that are seen as complementing these conventional strategies. The nodal methods we have outlined seek to enhance the bargaining power of weak actors in the contests that define the political terrain. These methods thus seek to promote what Drahos and Braithwaite (2002) have termed 'democratic bargaining'.

The idea here is to recognize that in cases where there are competing preferences, bargaining power counts. If poor constituencies are going to have a greater say in shaping governance, they need to develop greater bargaining power for their voices to be heard. Within a nodally governed world, effective nodal access is a prerequisite for bargaining.

Democracy, as we have come to understand it, means more than majority governance. It also means governance that is constrained by and seeks to accomplish outcomes that realize widely held values. There are many statements of what such values should be that have varying degrees of support. These are articulated in a variety of norms or codes by political, ethical and religious authorities. These codes act politically as resources that actors draw upon to justify and legitimize their courses of action. They also provide a basis that can be used to challenge actions of others that do not comply with codes. Securing nodal access that allows actors to engage in such challenges enables them to use these codes to their advantage.

Conclusion

In this chapter we have sought to provide guidance as to how weak actors might go about levelling the playing field of governance in ways that will accord them a greater voice, and through this, greater bargaining power. This idea of creating greater bargaining power for weak actors so that they might engage in democratic bargaining does not address the issue of how this bargaining might and should be regulated within a nodal world. It is to this question, of the 'governance of governance', that we now turn.

For a long time the issue of regulating the governance of security has been framed as a question of governing state institutions in order to ensure compliance – particularly on the part of police and military actors – with democratic standards of conduct. In a nodal world, this state-centred view is no longer sufficient. The issue of governing governance must now be framed in the context of a polycentric world of nodal governance. In the chapter to follow we outline our approach to this issue – an approach based on the assumption that the governance of governance must involve a mix of mechanisms, a set of hybrid instruments, that are pulled together in ways that harness the strengths of each while compensating for the weaknesses of the other (Lewis and Wood 2006).

Chapter 5

The governance of governance

Introduction

This chapter addresses the challenge of governing the range of nodes and nodal assemblages that now function to produce security goods across local, national and international levels. It is concerned, in short, with the 'governance of governance'.[1] Upon reading the previous chapter, some might argue that it is one thing to stress the importance of developing methods of power for weak actors to participate more meaningfully in the identification of security outcomes and in their provision. The normative agenda we sketched is one that accepts nodal governance not only as an empirical reality, but as a way of thinking that opens up new opportunities for realizing democratic values. It is another thing, however, to ensure that security delivery nodes (be they relatively strong or relatively weak) are themselves governed in accordance with normative and other standards of conduct.

In this chapter we explore the problem of 'governing governance' from a nodal perspective. We argue that this problem is best approached by pursuing a mix of forms or 'modalities' (Scott 2005) that, together, compensate for the weaknesses of each mechanism on its own. In other words, the answer rests in hybridity (Murray and Scott 2002; see also Goodin 2003; Lewis and Wood 2006). This hybridity is conceived not only in terms of diverse instruments or tools of governance, but also in terms of multiple governance authorities including (and very importantly) state authorities, as well as non-state authorities ranging from community-based groupings to NGOs to corporations.

In the first section of this chapter we examine developments in the governance of governance within the state realm, using the case of public policing as our main illustration. We examine the extent to which hybridity has already been pursued in the context of the 'new regulatory' state (Braithwaite 2000a) or 'regulatory capitalism' (Levi-Faur 2005). In public policing, as in other public sector contexts, attempts have been made to introduce new mechanisms for shaping the conduct of service providers to the end of maximizing efficiencies while driving ownership and steering to more localized levels. As part of this 'new management' (McLaughlin and Murji 1995, 1997), police organizations, and state agencies generally, have adopted the 'figurative logic' (Shearing and Ericson 1991) of business, seeing themselves as 'enterprises' devoted to 'customized service delivery' (Wood 2000). Based on research we conducted in the 1990s on regulatory trends in the province of Ontario, Canada (Wood 2000), we also observed a 'meta-regulatory' (Scott 2003; Parker 2002) or 'meta-governance' agenda devoted to taming marketization through the imposition of public service delivery standards.

The story of hybrid governance is not simply a story about the incorporation of market logics into the minds of public sector bureaucrats, and in particular the introduction of tools for results-based management. In other words, it is not simply the story about efforts on the part of states to allow the cross-pollination of ideas from the private sector to the public sector and back again. Hybridity in the governance of governance has also been accomplished through the activities of non-state authorities (Scott 2002), including NGOs, devoted to ensuring the compliance of corporate and state actors with particular normative standards, such as, for example, environmental standards (Gunningham *et al.* 2003) or human rights standards (Watchirs 2003). In the cosmopolitan governance of governance, referred to in Chapter 3, international rules are monitored and enforced by a variety of actors, through a variety of mechanisms, across a variety of micro and macro levels. In this way, it could be said that cosmopolitan governance is nodal governance within a meta-normative frame.

There is also much self-governance that takes place not only within public sector organizations like the police, but also in the case of commercial service providers such as private security organizations (Stenning 2000; on corporate self-governance see Parker 2002). Of course, each governance mechanism expresses its own mentality and each, on its own, is insufficient (see Goodin 2003), not only in guaranteeing full compliance on the part of targeted nodes, but also on the part of other nodes within the broader field of security provision. In this chapter, we do not provide a concrete proposal for an ideal hybrid model. Rather, we draw inspiration from broader developments in regulatory theory to identify avenues where enhanced hybridity, and hence greater robustness in the governance system, might be explored.

Hybridity in state governance: the case of public policing

Conceived in conventional terms, the governance of governance is a process carried out by state institutions in furtherance of objectives that are developed and established by state institutions. In the regulation literature, this process involves 'the attempt to modify the socially-valued behaviour of others by the promulgation and enforcement of systems of rules, typically by establishing an institutionally distinct regulator' (Loughlin and Scott 1997: 205). Over time, however, scholars of regulation have discovered the limitations of this rule-based conception. 'Socially-valued behaviour' is expressed in a variety of norms, and not necessarily norms that are codified in formal rules. Normative systems are plural, as are the mechanisms that promote compliance with such systems.

Scott offers a slightly broader conception of regulation as

> any process or set of processes by which norms are established, the behaviour of those subject to the norms monitored or fed back into the regime, and for which there are mechanisms for holding the behaviour of regulated actors within the acceptable limits of the regime (whether by enforcement action or by some other mechanism). (Scott 2001: 331 in Lange 2003: 411)

This somewhat broader conception allows for considerations of non-state actors as well as mechanisms for the governance of governance that do not rely on enforcement and the rationale of deterrence. It is this broad understanding of regulation that we conceive of as the 'governance of governance'.

Traditionally, the challenge of governing public policing has been approached according to the narrower understanding of regulation just discussed. Specifically, it has been explored in terms of questions of accountability and control. Accountability is a retrospective form of governance that involves scrutiny of decisions and past actions. It involves 'the duty to give accounts for one's actions to some other person or body' (Scott 2000: 40; see also Stenning 1995: 5), and is concerned with the following key issues:

> *What* type of decisions do the police make, explicitly and implicitly, in exercising their powers? *To whom* should they be accountable for the different sorts of decisions? *What type* of accountability should they have to the relevant bodies? *What mechanisms* should be established to deliver effectively the appropriate type of accountability to such bodies? (Reiner 1993: 6, italics in original)

Accountability can thus be seen as one subset of governance which more broadly includes a wide variety of backward-looking and forward-looking mechanisms for shaping conduct (Mulgan 2002: 4). The key mechanisms shaping the conduct of public policing organizations can be defined in terms of *legal accountability* and *political accountability*.

Legal accountability

As Parker and Braithwaite note, when public police organizations were established in different parts of the world, they 'came to rival and even surpass tax administrations as the largest regulatory bureaucracies' (Parker and Braithwaite 2003: 121). As such, they emerged as a central target of governance because of the potential threats they posed to public welfare: 'The powers that the police possess to protect fundamental liberties simultaneously provide the potential for severe abuse of these freedoms' (Jones 2003b: 606).

The potential for severe abuse of freedoms continues to reside in the capacity of the police to exercise or threaten violence on behalf of the state, although clearly this potential does not reside exclusively in the police, or the military, as we will discuss in the context of commercial security provision. Conventional approaches to the governance of the police have thus been largely concerned with shaping the ways in which police officers threaten and exercise coercion. Within a rules-based conception, the capacity of the police to exercise coercion needs to be constrained by the very laws that they are enforcing (see McBarnet 1979). Central to this constraint is a system of legal accountability, expressed concisely in the Patten Commission Report on Policing for Northern Ireland:

> The police are tasked to uphold and if necessary enforce the law, but, like any citizens, they must at all times act within it. Police officers should have sound knowledge of the law and of their powers under it. They need sufficient discretion to do their jobs well but they need at the same time to be monitored in their adherence to the law, and to have any errors rectified and abuses punished. It is important for the credibility of the police in the communities they serve that all this should not only be the case but that it should also be seen to be the case. (Patten 1999: 25)

The principal mechanism developed in established democracies to respond to the potential for, and the reality of, such abuses of power has been, and continues to be, transparency. Initially, the principal institutional arrangement used to provide for transparency was a system for complaining about police practice (Goldsmith 1991a). The focus of complaint systems has been on identifying, and then punishing, individual police officers who have violated the law and other rules. The arguments and debates surrounding complaints processes, which continue today, have concerned such matters as who should investigate complaints and who should decide on penalties for wrongdoing (Goldsmith 1991b, 1995, 2000).

Traditionally, complaints processes have not been organized to generate findings about broader administrative or policy issues (Goldsmith 1995: 116). Rather, they have been focused on the conduct of individual policing agents (Jones 2003a: 2). By personifying or individualizing issues, remedial measures often tend to be internal discipline undertaken by senior police officers, or in some cases criminal prosecution. Goldsmith writes:

> In this state of affairs, there is no admission of more pervasive organizational factors at work in explaining the action giving rise to the grievance, so that the domain of police administrators in relation to setting matters of policy emerges unscathed. Complaints viewed in this way become handmaidens of internal police discipline, rather than 'windows' into the organization. (Goldsmith 1995: 116)

As in other spheres of governance, this form of 'regulatory legalism' (Braithwaite 1993) reflects a dominant view of how compliance with norms of conduct can be achieved at the level of both individuals and organizations. This view assumes that behaviour can be shaped by 'command and control' mechanisms that work at the level of deterrence. Through practices of rule enforcement, individuals, groups or institutions that violate normative standards of conduct are punished according to relevant sets of legal and operational rules (see Chan 2001b: 140). Consistent with the 'rational' conception of human agency that has guided police governance of citizens (see our discussion in Chapter 2), the governance of the police enterprise itself is based on the assumption that non-compliance with rules will be avoided due to threats of punitive sanctions (see Bayley 1995: 94–6). Put another way, there is a preference for what May describes as the 'hard regulatory path' aimed at 'activating deterrent motivations' (May 2002: 5).

Within this governance framework, the most controversial debates surrounding legal accountability have centred on the question of whether, and to what degree, self-governance of police behaviour should be linked to external governance by non-police bodies (Goldsmith 1991a, 1991b), typically undertaken through external public complaints procedures which complement other mechanisms including criminal prosecutions and civil actions (Goldsmith 1995). The question of how to achieve the right balance between self-governance and governance by others has been central to such debates. In order to ensure more than simply 'token compliance' (Reiner 1993: 21), there must be trust in the police to self-govern responsibly and in accordance with their professional capacities and resources (Brown 1981; Goldsmith 1995: 115; Bayley 1983). If the balance is tipped too far in favour of external regulation, police organizations, as seen in other organizational contexts (Parker 2002) may react through forms of indifference or resistance to the regulatory process (Stenning 1995; Jones 2003b: 605).

Another critique that has characterized debate within the legal accountability framework has focused on the high levels of autonomy and discretion exercised by police officers. The decisions officers make are based on very specific and contextual human judgements (Lustgarten 1986), many of which are often unknown by senior officers and others who are not present at the time such decisions are made. This is especially the case when officers choose *not* to enforce the law when this option is available to them (Lustgarten 1986, ch. 1; Reiner 1993: 11). As such, the innumerable variables and contingent events shaping decision making, along with the considerable autonomy officers have to make decisions, combine to make it very difficult for senior managers and political authorities to establish formal rules of conduct apart from internal organizational procedures (Reiner 1993: 9).

Studies of police culture have revealed that the behaviour of officers is shaped by a wide range of forces and variables and that informal rules guide behaviour just as much as or more than formal rules (Manning 1977; Ericson 1981; Reuss-Ianni and Ianni 1983). Police culture, according to Goldsmith, plays a 'normative role ... which works by "filling in" areas of formal policy vacuum (what also might be described as areas of broad police discretion) with informal rules of thumb provided by the police officers themselves' (Goldsmith 2000: 125; see also Reiner 1993: 13). The 'recipe rules' (Ericson 1981) that shape police decision making, combined with the considerable autonomy of police officers on the ground, provide the conditions for what McBarnet (1979) calls 'creative compliance', which 'involves avoiding law's requirements without actually contravening them. It involves creative use of the "material" of law (McBarnet 1984) to construct devices which comply with the letter of the law, while nonetheless escaping from legal control' (McBarnet and Whelan 1997: 178).

If culture is understood as 'a system of shared values and understandings' (Chan *et al.* 2003: 3) or as 'a sensibility, a way of being out of which action will flow' (Shearing and Ericson 1991: 491) then governance itself will only be effective if this other, more informal normative system is altered. Therefore, it has been argued that governance must be conceptualized within a broader programme of cultural change, the dimensions of which would cut across the entire social and political 'field' of policing (Chan 1997, 2001b; Chan *et al.* 2003). This would include 'fundamentally altering our view of the police mandate and changing the rule systems by which they operate' (Ericson 1981: 102).

Political accountability

Political or democratic accountability, the second key mechanism for governing public policing, is concerned with providing public input into the general policy decisions guiding the provision of police services (Jones 2003a: 2). It is concerned with ensuring a certain 'steering' role (Osborne and Gaebler 1992) in the determination of police priorities. Two main factors have placed limits on this steering capability. First, the doctrine of 'operational independence' suggests that because of the unique expertise of the police, they themselves are in the best position to determine strategic priorities and allocate resources. The second factor relates to the principle of impartiality underlying policing in democratic societies. The problem raised is that elected governments are by definition partisan. They represent the interests of one party or a coalition of parties who in turn represent the interests of only a fraction of the citizens of a country. The concern is that placing the police under the direct control of governments is effectively to place them under the control of a partisan political party that represents only a fragment of the population. The debate around political accountability has been concerned with finding ways of avoiding this problem.

The principles of impartiality, autonomy and operational independence embody the assumption that the police must be 'apolitical' or 'above politics' (Goldsmith 1995: 115) in order to be legitimate. While historical struggles with police corruption and political influence belie this strong normative concern, policing, as many point out, is inherently political (Lustgarten 1986; Martin 1995; Reiner 1995). The determination of broad police priorities as well as choices about where and how to allocate police resources invariably privilege certain interests over others (Martin 1995: 142–4; Jones 2003b: 606).

The most widespread solution is one that claims that while general policy matters are ones that governments should be able to control, the day-to-day actions of the police should be independent of political interference. It is this very argument that results in an emphasis on legalistic governance discussed above. In place of strong steering, involving direct control of police strategies and operations, control is exercised indirectly through complaint systems. The logic expressed is that the police should be accountable to the law and the law alone. This conception of police independence was expressed by Lord Denning in the commonly cited English case of *R. v.* Metropolitan Police Commissioner ex parte Blackburn where he remarked that 'he [the Commissioner of the London Metropolitan Police] is not the servant of anyone, save of the law itself' (*R. v.* Metropolitan Police Commissioner, ex parte Blackburn, [1968] 1 All E.R. 763, at 769 – *per* Lord Denning, M.R.).

Despite the above critiques of traditional police governance, the system as a whole has become more robust over time in Australia, Canada and other established democracies. There are nodal arrangements for undertaking 'vertical' accountability – involving government agencies, parliaments and courts – as well as 'horizontal' accountability achieved through measures such as freedom of information and judicial review. Such external governance arrangements are complemented by self-governance processes including application of police internal rules, regulations and codes of conduct, the operation of internal investigation units as well as establishment of internal 'ethical standards' units whose existence was prompted, in the Australian context, by a series of scandals related to systemic instances of police corruption and abuse of power (Lewis and Wood 2006).

The question of whether this existing accountability framework will prove sufficiently robust, or whether aspects of it will 'lose their teeth' in the present context of the 'war on terror' is a matter of considerable controversy. In the case of Australia, the public police have been granted significant new powers through the new Anti-Terrorism Act 2005 (Commonwealth) which include, for example (depending on the jurisdiction), preventive detention for up to fourteen days and covert search warrants that need only be approved by a member of the executive (Lewis and Wood 2006). While such powers have been granted in the name of Australian homeland security (2006), some suggest that the human security (particularly the human rights and freedoms) of suspects or otherwise innocent individuals could be trumped through the anti-terrorism agenda (see generally Daniels *et al.* 2001 for debates in the Canadian context). Whether this tension between homeland security and human security can be addressed through new forms of political accountability or through other governance mechanisms is something we consider towards the end of this chapter.

The new regulatory state or regulatory capitalism

While debates surrounding the improvement of legal and political accountability of public police continue to this day and continue to lead to corrective measures, there has been a different kind of shift in public sector regulation in the form of the 'new regulatory state' (Braithwaite 2000a) or what Levi-Faur (2005) prefers to call 'regulatory capitalism'. A central feature of regulatory capitalism has been the incorporation of business-like thinking into the public sector, resulting in an entrepreneurial ethic in the delivery of services. This ethic has led to institutional restructuring and corporate 'reengineering' programmes (see Hammer and Champy 1993) aimed at enhancing efficiencies. It has also led to an uncoupling of 'steering' and 'rowing' (Osborne and Gaebler 1992) functions manifested in privatization and through the creation of market-based environments, which has required public agencies to compete with other state or non-state providers. This marketization process has not been accompanied by deregulation, but rather by an enhancement of regulation (the governance of governance) on the part of the state or its delegated regulatory agencies. Thus, there has been a rise in new regulatory mechanisms targeted at both state and non-state providers (Loughlin and Scott 1997: 212–3; Parker and Braithwaite 2003; Jordana and Levi-Faur 2004). With this in mind, Levi-Faur points out that there is more to the story of regulatory capitalism than simply the enhancement of regulation by the state: 'More regulation is observed not only in the context of the traditional relations between governments and business but also within the state, within corporations, and in social (not only economic) arenas' (Levi-Faur 2005: 20). This is a point that we develop throughout our analysis.

Thinking like a business

If we return to the public policing context, we find, particularly during the 1990s, a way of thinking that established new kinds of governance relationships between service providers and their clients (i.e. 'customers') and between the police and the formal authorities to whom they were accountable. In our Canadian study, the Ontario Provincial Police (OPP) had been experiencing, during the 1990s, 'increased financial pressures and public demand for greater effectiveness, efficiency and accountability in the delivery of public services' (Ontario Provincial Police 1993: 14). The OPP claimed that they would have to provide 'better customer service at reduced levels of spending', while being less 'bureaucratic, more flexible and responsive to the needs of the public' (Ontario Provincial Police 1995b: iv). While it was, and still is, not the objective of the OPP to make a profit, police managers opted for market disciplines (Jones 2003b: 616) as a means of cultivating an 'anti-bureaucratic ethos' (De Lint 1998) and appealing to the 'product producing organization – assumptively a corporate and for-profit organization – as the root unit in society' (De Lint 1998: 263; O'Malley 1997).

Thinking like a business has involved the establishment of contractual or quasi-contractual relationships between sponsors and providers of services. In the case of public policing, 'sponsors' come in two forms: those who pay for policing services directly out of taxes they pay (citizens), and the government agency/ministry that administers the policing budget on behalf of taxpayers. As part of the new public management, the 'policing ministry' (to use a generic term) seeks to forge the conditions in which citizens contract to the public police in the capacity of 'customers' who are receiving clearly defined products and services.

In constituting a customer relationship, service delivery terms are spelled out, including specification of the nature of the services and products that are to be 'purchased' as well as the cost of each. In the province of Ontario, budgets for local policing services and products were shifted as much as possible to local levels. This served to more clearly delineate lines of authority and fiscal accountability, as well as reduce 'free riding', where some consumers were, it was argued, benefiting from goods they did not directly pay for. A major issue for the former Conservative Party administration in Ontario during the heyday of neo-liberal reforms in the 1990s was that some municipalities were required to pay for their security provision through municipal taxes while others received the services of the OPP at no cost due to subsidies by the province. This inequity had been identified approximately 25 years prior in the 1974 Task Force Report on Policing in Ontario, when it recommended that

> the structure of policing in Ontario be realigned so that each community will either provide its own policing or obtain police services through a negotiated contract with the Ontario Provincial Police or other operating force. This rationalization of police forces in Ontario should eliminate the problem of free policing. (Task Force on Policing in Ontario 1974: 125)

Over 20 years later, when the Ministry revisited this issue, it decided that the 576 'free riding' municipalities should pay and be responsible for their policing services 'regardless of the type of arrangement for police services' (Ministry of Solicitor General and Correctional Services 1996: 6; see also Wood 2000). With 'equitable financing', the municipalities that received OPP services, but without a contract, were now required to determine the nature of their service delivery and to pay for it (Police Services Act, R.S.O. 1990, c. P-15, s. 5.1). In this 'quasi-market' (Johnston 1999: 181) of rural and small-town Ontario, the OPP set out to become 'the community police service of choice outside large urban areas' (Ontario Provincial Police 1997: 16). Today, the OPP has effectively remained a strong player in the policing market outside large urban areas, partly because they have an economy of scale and provide specialized services, and partly because of the 'symbolic capital' (Dupont 2006) they possess in Ontario communities.

The injection of a business ethos in Ontario governance as well as with similar neo-liberal reforms of the late twentieth century elsewhere did not involve a turn to unfettered market relations, but rather involved measures to 'constitute' market relationships (see Shearing 1993) in ways that promoted public interest objectives. One means of doing so was to establish standards of service delivery and to engage in 'meta-monitoring' (Grabosky 1995: 543) of those standards.

In Ontario, efforts to enhance local sponsorship of policing were accompanied by the establishment of province-wide standards of adequacy. The Regulation titled Adequacy and Effectiveness of Police Services, which came into effect in the year 2000 (O. Reg. 3/99 made under the Police Services Act, R.S.O. 1990, c. P-15), marked 'for the first time … standardized, province-wide effectiveness measures that allow the province, local governments, police services boards and the community to assess the value and impact of their investment in policing' (Government of Ontario 1998). Underlying such standards is a decisive list of core functions – described as 'crime prevention', 'law enforcement', 'victims assistance', 'public order maintenance', and 'emergency response services' (O. Reg 3/99) – that every municipal police service provider would be required to carry out in order to achieve a provincial standard of adequacy (Government of Ontario 1998).

The Provincial Adequacy Standards were applied in tandem with six service delivery options listed in the Police Services Act:

1 The council may establish a police force, the members of which shall be appointed by the [police services] board …

2 The council may enter into an agreement … with one or more other councils to constitute a joint board and the joint board may appoint the members of a police force …

3 The council may enter into an agreement … with one or more other councils to amalgamate their police forces.

4 The council may enter into an agreement … with the council of another municipality to have its police services provided by the board of the other municipality … if the municipality that is to receive the police services is contiguous to the municipality that is to provide the police services or is contiguous to any other municipality that receives police services from the same municipality.

5 The council may enter into an agreement … alone or jointly with one or more other councils, to have police services provided by the Ontario Provincial Police.

6 With the Commission's approval, the council may adopt a different method of providing police services … (Police Services Act, R.S.O. 1990, c. P-15, s. 5. (1))

These options are meant to provide municipalities with flexibility in determining how best to meet the Provincial Adequacy Standards, including the development of tailor-made security contracts with service providers that balance service levels with costs. Contractual relationships with service providers other than the public police are encouraged, but within a meta-governance arrangement where the public police force monitors such services. For example, in meeting the core service delivery requirement of 'crime prevention', the Regulation states that the Police Services Board may 'enter into an agreement with one or more organizations other than police forces to have the organization or organizations provide crime prevention initiatives *under the direction of a member of the police force*' (O. Reg. 3/99, s. 1. (2) (b), italics added). It further stipulates:

> *Every chief of police shall establish procedures and processes* on problem-oriented policing and crime prevention initiatives, whether the police force provides community-based crime prevention initiatives or whether crime prevention initiatives are provided by another police force or on a combined or regional or co-operative basis or by another organization' (O. Reg. 3/99, s. 3, italics added)

The same emphasis on oversight by police is articulated in other sections of the Regulation, such as in the monitoring of communications and dispatch services or crime analysis specialists where such services are not provided by police force personnel (O. Reg. 3/99, s. 5-7).

Service delivery standards not only establish public interest norms for the delivery of policing services, but provide the framework within which 'auditable' criteria of performance can be established. The regulation discussed above requires Police Services Boards to establish a 'business plan' for its service provider at least once every three years, which must cover issues such as 'core business and functions of the police force', as well as 'quantitative and qualitative performance objectives and indicators' in areas including 'the police force's provision of community-based crime prevention initiatives, community patrol and criminal investigation services', and 'community satisfaction with police services' (O. Reg. 3/99, s.

30). Police organizations are required to submit a report to the Board per fiscal year which must include information on their performance against projected indicators/targets (O. Reg. 3/99: s. 31).

Police organizations across the globe are now compelled to prepare business plans. In contrast to the individualistic regulation of police officer conduct, business plans provide a tool for the meta-monitoring of police organizational performance as a whole (Auditor-General's Office 2004: 2). Business plans establish 'outcomes' (products and services) as well as measures (performance indicators) that provide an index of progress towards outcomes. As discussed above, the clear delineation of outcomes provides an essential basis for purchaser–provider relationships. It must be clear what services and/or products the customer is expecting as well as the cost of each product.

While business plans are imagined as a contract with citizens, their specific outcomes are aligned with the broader outcomes dictated by higher levels of government (e.g. provincial or state levels). In the case of Victoria Police in Australia, their performance outcomes are aligned with the state-wide normative agenda articulated in a report titled *Growing Victoria Together*, which focuses on government services that 'build cohesive communities' and 'reduce inequities' (Department of Premier and Cabinet 2001; Victoria Police 2003: 7). Thus, while business planning allows for some flexibility in the determination of local service delivery priorities, the contract with citizens implied by this process is aligned with public interest imperatives that are determined at higher levels of government.

This alignment of desired outcomes across levels of government reflects a form of 'action at a distance' which we discussed in previous chapters (Latour 1987; Rose and Miller 1992). Those government departments and ministries that function as auspices of police service delivery can steer the provision of services while remaining at a distance from specific substantive and operational issues. Business planning is a 'technology of distance' (Espeland 1998: 1107), a key component of which is quantification. In this there is 'trust in numbers', where emphasis is placed on 'mechanical objectivity' rather than exclusively on the discretion of experts (Porter 1995). In contrast to tangible products, like security hardware, 'social' outcomes (e.g. perceptions of safety, good customer service, 'real' levels of security) in business plans require indicators that establish links between particular inputs (e.g. increased number of police patrols) with such outcomes.

Linking inputs to outcomes raises the complex problem of 'attribution' – that is, how to determine whether particular actions produce certain effects, especially when there is a multitude of variables that influence an outcome (Mayne 1999). In the case of policing outcomes, where attribution is inherently problematic, service providers commit to outcomes that in their judgement are something they can have reasonable influence over. In other words, expected outcomes must be 'realistic', and based on an assessment of what one's capacities and resources are capable of delivering (Auditor General of Canada 2002: 9). In emphasizing ends that can be controlled, the new public management supports a cautious optimism about the contribution of police inputs to various social outcomes. This is evidenced in the business plans that are now prepared by police organizations in different parts of the world. Consider a recent business plan of Victoria Police in Australia that established the following key targets:

- Reduction in the crime rate by 5%.
- Reduction in the road toll and incidence of road trauma by 20%.
- Increase levels of community perceptions of safety by 1.5%.
- Increase levels of customer satisfaction with policing services by 2.6%. (Victoria Police 2003: 6)

Targets such as these allow for 'commensuration', which is 'the comparison of different entities according to a common metric' (Espeland and Stevens 1998: 313). Commensuration allows consumers of policing services and products to compare the performance of different police organizations. For ease of comparison, common features of service delivery across different service providers must be isolated and measured similarly according to easily quantifiable indicators. In Australia, the 2004 report of the Steering Committee for the Review of Government Service Provision (SCRGSP), published by the Productivity Commission, compared the performance of all state police services. The Committee identified four core areas (outcomes) of police service delivery for which it established performance indicators. These 'service delivery areas' (SDAs) are:

- To allow people to undertake their lawful pursuits confidently and safely (through activities associated with *community safety and support*).
- To bring to justice those people responsible for committing an offence (through activities associated with *crime investigation*).
- To promote safer behaviour on roads (through activities associated *with road safety and traffic management*).

- To support the judicial process to achieve efficient and effective court case management and judicial processing, while providing safe custody for alleged offenders, and ensuring fair and equitable treatment of both victims and alleged offenders (through activities associated with *services to the judicial process*). (Steering Committee for the Review of Government Service Provision 2004: 5.16; italics in original)

The text in parentheses reveals the 'inputs' that the Steering Committee expects will produce the four key results. Performance indicators are established for each input, against which each state police service is assessed in relation to one another as well as in relation to the national average. The essence of the Steering Committee's report is a series of tables for comparing performance across states along lines such as 'people who were "satisfied" or "very satisfied" with police services' (Steering Committee for the Review of Government Service Provision 2004: 5.19) and 'people who "agreed" or "strongly agreed" that police treat people fairly and equally' (2004: 5.23). Such measures of customer satisfaction capture 'process' outcomes with respect to the quality of the interaction between police and clients. In addition to such 'perceptions' indicators, states are compared across 'actual' levels of recorded crime, such as 'recorded victims of murder' (2004: 5.35), and 'recorded victims of motor vehicle theft' (2004: 5.38).

The new managerialist strand of output-based performance measurement has been taken up in earnest in both the 'broken windows' and 'intelligence-led' waves of community policing. Bratton, during his 'reengineering' of the New York Police, focused on enhancing accountability for 'results'. The implementation of the Compstat programme for monitoring crime statistics reflected the view that crime control can be organized as a business enterprise (Logan 1999: 344). Programmes under the banner of 'intelligence-led policing' are similarly oriented towards enhancing accountability for the 'fundamental crime control mission' (Weisburd *et al.* 2003: 426). Technological mechanisms of accountability such as Compstat provide the capability for 'strategic problem solving' (2003: 425) of the kind advocated by Goldstein – that is, the use of data to systemically analyse the problems that come to the attention of the police. With crime statistics as the essential outcome measure of police performance, the elimination or minimization of residual social problems emerge largely as a means for preventing crime, rather than an end in and of itself.

This shift in emphasis towards outcome-based accountability in policing, and for 'crime control' in particular, has not been uncontested and indeed is seen by some as having generated new challenges for the governance of governance. The case has been made that process standards have been weakened at the expense of outcome standards, a case that continues to be put forward in light of the war on terror and the increased police powers that have come with it in a variety of countries (Daniels *et al.* 2001). In the United States, some rather 'hard' applications of the broken windows logic, and the Compstat system that supports it, have led to major controversy surrounding allegations of police abuse in the New York Police Department. For example, Amadou Diallo was shot 19 times by four officers who had released a total of 41 bullets in response to Diallo allegedly 'acting suspiciously' and matching the description of a rape suspect (Harring 2000: 9–10).

Harring argues that this incident was the product of a regular routine of 'roustings' that are encouraged by Compstat. This mechanism of 'computer-based police accountability' calls for the production of statistics concerning the 'stop' and 'search' of individuals who look likely to be carrying guns or contraband (on the contested nature of 'reasonable' police searches see Logan 1999: 346–56). Harring argues:

> One after another, young men are confronted, ordered to stop, searched, verbally engaged for information about themselves or others, and then either let go or arrested on petty charges. The squads then move on from one roust to another, all shift long, producing the desired arrest statistics that underpin the functioning of the policing machine. (Harring 2000: 10; see also Logan 1999: 345)

According to Steiker, this practice reflects one of a variety of 'prophylactic measures' associated with the 'Preventative State' (Steiker 1998 cited in Logan 1999: 345).

We revisit this normative concern and related others towards the end of this chapter as we consider a possible agenda for assessing and addressing existing mixes of governance mechanisms and authorities. Our explanatory point is simply that the public sector reforms that have characterized the 'new regulatory state' in Western democracies have involved a melding or mixing of state and market logics for the governance of governance. In the following section we explore the theme of hybridity of governance mechanisms and authorities beyond the state, where both the functions of steering and rowing have diversified. Parker and Braithwaite point out that 'in the new regulatory state, not only does the state do less rowing and more steering, it also does its steering in a way that is mindful of a lot of steering that is also being done by business organizations, NGOs, and others' (Parker and Braithwaite 2003: 126). In this section we examine forms of steering undertaken by state and non-state auspices in relation to the conduct of both state and non-state providers of goods. In our discussion we draw from developments not only in the governance of public and private policing, but also in other realms, more closely associated with the provision of human security, that have been the focus of regulatory scholars more generally.

Hybridity in decentred governance: private policing and beyond

As we have argued previously, the rise of neo-liberalism and the new regulatory state only serves as a partial explanation for the growth in private governance authorities and providers. The complexity of 'decentred' regulation (Black 2000) is exemplified by developments in private policing (Shearing and Wood 2000, 2003b) and arguably more so in domains beyond security (as traditionally conceived). As we discussed in Chapter 1, private security has developed, in large part, because corporations have designed strategies for governing security that allow them to have a greater say in governance than they would have if they were to rely on state-based conceptions of what counts as security and how it is to be governed. In contrast to privatized services like telecommunications or electricity or even prisons, the growth in private policing has not been the direct outcome of neo-liberalism. While states engage in 'contracting out' services to private providers, forms of 'private government' (Macaulay 1986) have also proliferated in spaces and places protected by property laws or in transnational realms that span the sovereign jurisdictions of nation-states (Hall and Biersteker 2002b).

The governance of private or commercial policing consists of a variety of mechanisms and a range of authorities. For instance, in the governance of private policing, and in particular 'contract' private policing, Stenning (2000) points to the 'regulatory pluralism' (Parker and Braithwaite 2003: 129) that is largely obscured from views that privilege the accountability models of public policing. For instance, depending on the jurisdiction, governance mechanisms include a mix of instruments such as licensing, registration, training requirements, and rules concerning insurance coverage. There are also forms of self-regulation aimed at achieving similar goals to state regulation. Private police are subject to criminal and civil liability. Labour or employment law can be utilized in the monitoring and enforcement of workplace standards in private police organizations. There is also liability for breach of contract, which is an important mechanism in 'contract' versus 'in-house' security arrangements. Insurance contracts stipulate requirements pertaining to security provision on insured properties. Moreover, in some jurisdictions minimum industry standards are established collaboratively by various stakeholder groupings. Finally, there is 'marketplace accountability', which is based on the simple premise that customers will take their business elsewhere if they are unsatisfied with the integrity and/or quality of the services they have received (Stenning 2000).

As with the governance of public policing, the governance of private policing has not been watertight. For instance, there are worries about the invasion of privacy that occurs on the part of 'private governments', such as employers who have the authority to access employees' computers (Joh 2005: 604). In Australia, private security guards have been involved in a range of problematic behaviours, from fraudulent conduct to the illegal access of information to incidents of shooting causing injury and death (Prenzler and Sarre 1998, 1999; Prenzler and King 2002).

Similar concerns are increasingly being raised with respect to the operation of the transnational security market. As Stenning suggested in the domestic context, 'it is now almost impossible to identify any function or responsibility of the public police which is not, somewhere and under some circumstances, assumed and performed by private police in democratic societies' (Stenning 2000: 328). A similar statement could be made with respect to the transnational realm where the nature of services and products on the security market is amazingly wide-ranging, from contract security companies to risk-management services to business intelligence to military services (Johnston 2006b: 36). What is particularly notable about the transnational context is that violence itself has been clearly marketized (Avant 2005) to the extent that private actors exercise coercion on behalf of state and/or non-state interests. While some private military firms provide various kinds of support services, such as logistical and technical support or strategic advice, some assist at the tactical level and are required to exercise some element of coercion in this capacity (Singer 2001, 2003).

While Stenning's statement about the scope of private security functions applies to established democracies, transnational security provision, by its very nature, spans countries in various stages of transition or democratic development. And of course, reports such as that released by the Human Security Centre (2005) clearly reveal that analyses of the 'formal' global security industry do not adequately capture all that goes on in the 'new wars', particularly with respect to the orchestration and operation of violence on the part of state and non-state actors that is not sponsored simply by states or even by corporations (Duffield 2002: 157–8). Indeed, as we discussed in Chapter 3, the global war economy (Kaldor 1999; Duffield 2002) is one that cuts across and ties together at different points the operations of both formal and informal markets. This means that it is difficult to establish clean lines around the governance of 'legitimate' security provision versus the governance of security provision that derives its authority and sponsorship from entities that might be advancing, in Kaldor's terms, a 'particularistic' normative agenda.

If we return for a moment to the formal 'legitimate' security market, we discover an array of challenges for the governance of governance (Stenning 2000: 338–9; Singer 2001, 2003; Avant 2005; Johnston 2006b), particularly if we frame our concerns within a state-centred model. As it stands we know that states, by virtue of not being the sole sponsors of transnational security provision, are not the sole governance actors. Rather, a range of other authorities, such as corporations and international organizations, sponsor and hence govern the entities that provide services and products on their behalf (Singer 2001). In the transnational security market 'neither supply nor demand fit within state borders' (Avant 2005: 144). As such, the state-based governance regimes that do exist have proven inadequate in addressing the multi-jurisdictional nature of service provision. By virtue of its global nature, this market provides the conditions in which security companies can outmanoeuvre state-based governance regimes (Stenning 2000: 338–9). As Avant describes it,

> the industry's low capitalization, fluid structure, and the lack of commitment to territory – a PSC [Private Security Company] frustrated with one state's regulation can simply move abroad, or melt and reconstitute itself differently to avoid it – decrease the usefulness of the kinds of authoritative controls often associated with states. (Avant 2005: 144)

With the mechanisms of 'contractual governance' (Crawford 2003) that do exist, a series of issues has been raised with respect to the monitoring of service delivery quality. For example, the very process of awarding contracts in the first place may be an unduly political process. For instance, in the United States context, effective lobbying of the Pentagon by or on behalf of particular security companies appears to enhance the probability of such companies being awarded a contract (Johnston 2006b: 42).

The quality of private military personnel has also been challenged. Staff from DynCorp, for example, were accused of gross impropriety linked to their alleged involvement in a prostitution and rape scandal (2006: 45). Johnston also refers to concerns about the quality of private operations including

> lack of uniform rules of engagement; complaints from some guards about being put into combat situations without adequate weaponry, training or equipment; and reports of poor communication links with military commanders, where security guards have been stranded and left without reinforcements when under attack. (Johnston 2006: 45)

The list of governance issues continues, including the concern that there has been a 'fudging' of roles between private personnel hired as 'guards' and those hired as 'soldiers'. While guards are not to engage in offensive action, they can experience heightened levels of risk. For example, insurgents tend to equate guards with combat troops, a problem that is exacerbated by the similar dress and appearance of these 'civil' and 'military' agents (2006: 45).

The 'marketplace accountability' that Stenning discusses in the domestic context is also a feature of global governance. But its limitation as a mechanism has been challenged. Consumers, including state governments, have considerable potential to shape corporate behaviour by specifying the kinds of outcome-level and process-level standards they require in the services they receive (Avant 2005: 144–5). Theoretically, this means that consumers can take their business elsewhere if such standards are not met in the first instance. While corporate reputation is clearly a driver of business decisions, Singer injects an element of cynicism regarding the potential for private military firms (PMFs) to breach their contractual obligations in situations of heightened risk:

> A PMF may have no compunction about suspending its contract if a situation becomes too risky in either financial or physical

> terms. Because they are typically based elsewhere, and in the absence of applicable international laws to enforce compliance, PMFs face no real risk of punishment if they or their employees defect from their contractual obligations. Industry advocates dismiss these claims by noting that firms failing to fulfil the terms of their contracts would sully their reputation, thus hurting their chances of obtaining future contracts. Nevertheless, there are a number of situations in which short-term considerations could prevail over long-term market punishment (Singer 2001: 205–6)

In the domestic context, states have been addressing some of the regulatory deficits associated with the private security market through rather traditional measures of standard-setting and monitoring. In Victoria, Australia, for example, new regulations set out in the Private Security Act (2004), which came into effect in July 2005, are aimed at raising industry standards, enhancing service delivery quality and bolstering the legitimacy of the industry in the eyes of the public. Measures include mandatory licensing for security guards, crowd controllers, investigators and bodyguards, and the mandatory registration of individuals or businesses operating as security advisors or installers of security equipment (Lewis and Wood 2006).

These types of measures reflect current attempts on the part of states to enhance the social regulation of markets rather than move away from the marketization of security itself (Lewis and Wood 2006). As the regulatory literature has shown for some time now, the future of governance certainly does not rest on a contest between state-based and market-based solutions or between regulation and deregulation (Grabosky and Braithwaite 1993). Rather, scholars have been exploring the kinds of governance mixes that have been working well, or that could work well in a variety of empirical domains. Gunningham and Grabosky use the term 'smart regulation' to denote the potential for improving such mixes. In the context of environmental regulation, their argument is that 'recruiting a range of regulatory actors to implement complementary combinations of policy instruments tailored to specific environmental goals and circumstances, will produce more effective and efficient policy outcomes' (Gunningham and Grabosky 1998: 15). Their emphasis is on 'the potential for second and third parties (business or commercial or non-commercial third parties) to act as surrogate or quasi-regulators, complementing or replacing government regulation in certain circumstances' (1998: 15).

The potential for smart regulation can be seen in a variety of empirical domains where non-commercial third parties are shaping the conduct of both state and non-state actors. If we return to our example of human rights activism in Argentina discussed in Chapter 3, we find a mix of mechanisms being utilized that involves monitoring and exposure as well as command-and-control governance in the form of prosecuting those who have perpetrated state institutional violence. The governance 'authorities' in this case are not simply public prosecutors, but individual human rights activists, often working *pro bono*, local human rights organizations and national and international human rights bodies, such as the Inter American Human Rights Court and Commission (Wood and Font in press). Human rights organizations and research centres in Argentina also contribute to the development and articulation of policing and security proposals. An extensive network of such bodies recently released a report titled *Más derechos, más seguridad, más seguridad, más derechos: Políticas públicas y seguridad en una sociedad democrática* ('More rights, more security, more security, more rights: Public policies and security in a democratic society') (Centro de Estudios Legales y Sociales 2004). This document is aimed at promoting alternative approaches to dealing with crime and delinquency that are lawful and respective of human rights and as such are serving to advance human security and cosmopolitan norms.

Scott argues more broadly: 'Regulation of public sector bodies by non-state organizations is an important but neglected aspect of contemporary governance arrangements' (Scott 2002: 56). Non-governmental organizations or non-commercial third parties operate as effective regulators of corporate behaviour in a variety of domains. Through a range of mechanisms they are shaping the 'enlightened self-interest of individuals and corporations' (Levi-Faur 2005: 21–2) in a variety of empirical domains. This is particularly evident in the 'corporate social responsibility' movement whereby social justice and environmental protection concerns inform daily business practices (Courville 2003; Levi-Faur 2005: 21).

A significant feature of the governance activities of non-state authorities is the use of established mechanisms, like audit (Power 1997), certification and licensing – traditionally associated with public sector governance and/or business regulation – towards new, often 'nonfinancial ends' (Scott 2003). Watchirs (2003), for example, reports on an Australian pilot for a 'human rights auditing' of the state. Its emphasis is on 'bridg[ing] international obligations to national practice by objectively evaluating the extent of legal implementation of the

HIV/AIDS and Human Rights, International Guidelines' (2003: 244–5). The role of civil society (as represented by NGOs) in the auditing process reflects the attempt to 'decentralize accountability by wresting control of the human rights "non-agenda" from the government, which has diluted the scope and depth of human rights discourse' (2003: 45). Watchirs adds: 'Human rights auditing turns "new public management" principles and rituals, involving accountability tools measuring efficiency and effectiveness of performance, on government itself' (2003: 45). In this example, we thus find a different kind of hybridity – with respect to both governance mechanisms and auspices – from what we discussed earlier in relation to public sector governance reform.

In the area of sustainable agriculture, we find the use of certification mechanisms oriented towards the production of social justice and environmental norms. Courville's work explores the regulatory significance of organizations like the Fairtrade Labelling Organizations International (FLO), a transnational body that operates a certification and labelling programme identifying fairly traded products (e.g. coffee, honey, bananas) for consumers (Courville 2003a). Another example is the International Social and Environmental Accreditation and Labelling Alliance (ISEAL) that undertakes initiatives consisting of 'private, voluntary market-based certification and accreditations systems that provide sanctions and rewards for socially and environmentally preferable production and trade across international boundaries' (Courville 2006). The significance of such governance processes, according to Courville, is what they signify in terms of the central involvement of 'global civil society' (see Kaldor 2003):

> Concerned by the lack of effective governmental action across a range of social justice and environmental issues – from poverty alleviation in developing countries to the unrelenting destruction of critical habitat for endangered species, and from clear inequalities of international trade rules to the lack of enforcement of national labour and environmental laws leading to corporate abuses of human rights and environmental destruction – individuals, networks and organizations outside the realm of the nation-state have begun to take a more active role in processes of social change. These individuals, networks, and organizations comprise global civil society, where people interact across borders outside their identification with a specific state. (Courville 2006: 272)

The concept of 'licensing' has also been reconceptualized in promoting the normative objectives of civil society groupings. Gunningham, Kagan and Thornton (2003, 2004) provide new insights into why some corporations actually go beyond their compliance requirements with respect to legal regulatory standards of environmental protection. According to these authors, the tendency for corporations to 'over comply' cannot be explained by a 'command and control' emphasis on the rationale of deterrence. Corporations care a great deal about the social expectations around environmental protection and sustainability placed on them by community groups and non-governmental organizations.

In their study of pulp and paper mills in the context of the water pollution issue, these authors found that such organizations possess a kind of 'social license' representing 'the demands on and expectations for a business enterprise that emerge from neighbourhoods, environmental groups, community members, and other elements of the surrounding civil society' (Gunningham *et al.* 2004: 308). There is a variety of mechanisms that civil society groupings can use to shape the behaviour of corporations, including activities geared at the 'reputational' level of corporations. Not unlike the example of human rights NGOs in Argentina, such groupings can participate in campaigns for 'naming' and 'shaming' organizations whose business activities threaten the health of human beings or that of the environment. The cost of not complying could be the loss of 'reputation capital' (2004: 320–1).

In this example, 'smart' regulation of the environment is advanced through regulatory mix. Social licence processes interact with other more traditional regulatory (legal) and economic processes to shape the behaviour of corporate actors. For instance, the informal sanctions of the public and media can 'lever up' formal governance instruments by calling for a tightening of legal requirements that are seen as too broad and permissive. Similarly, sufficient community activism can prompt state regulators to act earlier and more frequently. The activities of 'social licensors' can generate significant economic repercussions – such as loss of sales and even a reduction in share prices or diminished access to capital – through negative publicity campaigns and protests. Consumer boycotts are typical of this kind of approach and can be quite effective depending on the extent to which they shape the opinions and preferences of consumers as well as purchasers of, in this case, pulp products.

As one final example of nodal regulation beyond the state we turn to the example of 'non-state market-driven governance systems'

(NSMD) elucidated by Cashore and colleagues (Cashore 2002; Cashore *et al.* 2004; Bernstein and Cashore 2005). Based on an understanding of consumer preferences, they argue that different private authorities can manipulate markets in ways that result in the production and distribution of goods that comply with norms in areas such as social and environmental responsibility. The central mechanism for inducing compliance in this setting is market incentives (Cashore 2002: 504). Cashore argues that the motivations for regulated actors to comply with the standards set by through NSMD are not simply traced to matters of self-interest and desires for profitability. Rather, '[t]he market provides the context within which material and short-term self-interest motivations intersect with moral and cognitive elements, which together determine whether and how different NSMD governance systems gain authority to make rules' (Cashore 2002: 505). NSMD systems work through processes that serve to align the complex motivations of the numerous actors that participate in markets, ranging from companies producing goods, to distributors of goods, to consumers (organizations and individual members of civil society) to non-governmental bodies such as environmental groups. In essence, the content of the rules that characterize and underpin NSMD governance are shaped by the interaction of the above actors as they seek to promote their material and moral interests (2002: 505).

In illustrating the magnitude of NSMD governance, Cashore draws on the case of forestry certification. Frustrated by existing attempts at sustainable forestry, environmental NGOs came together to form a private governance regime, in Drahos' terms a supra-structural node (Drahos 2005a) – the Forest Stewardship Council (FSC) – that set out to certify forest landowners and forest companies who practised sustainable forestry. The normative standards established for this regime pertained to performance-level measures in such areas ranging from workers' rights to environmental impact of forestry practices to preservation of old growth forests (Cashore 2002: 507). The FSC provides the 'carrot and stick' approach. The 'carrot' – the incentive to practise sustainable forestry – is the status and legitimacy of being a 'certified' company, while the 'stick' is wielded through activities like boycott campaigns of non-certified bodies (Cashore 2002: 507). Cashore and colleagues describe the 'non-state market-driven' character of these governance arrangements:

> [R]ule-making clout does not come from traditional Westphalian state-centered sovereign authority but rather from companies

> along the market's supply chain, who make procedures of these private governance systems. Environmental groups and other non-governmental organizations (NGOs) attempt to influence company evaluations through economic carrots (the promise of market access or potential price premiums) and sticks (public and market campaigns aimed at pressuring companies to support certification). (Cashore *et al.* 2004: 4)

A core theme cutting across the above examples of nodal, hybrid governance of governance is a concerted effort not to undermine the operation of either states or markets but rather to shape the flow of events in both state and market-based activities in ways that contribute to outcomes in areas such as social justice, sustainable environments and agriculture, and the protection of human rights, areas that all link to the human security vision depicted in Chapter 3. States have always blended economic and social regulation, but in this global, nodal world in which we live it is no longer sensible, we argue, to assume that states should be the ones (assuming they possess the knowledge and capacity) to monopolize the governance of governance. Indeed, in some contexts, states themselves are laggards, or even outright violators, with respect to compliance with some normative standards. We now turn briefly to the issue of future directions in the crafting of nodal governance regimes based on mixes of mechanisms and authorities.

Nodal governance for the future

With respect to the future of governance in a nodal world, we begin with a quote from Johnston:

> It is one thing to invoke good normative reasons to justify why the state *should* exercise meta-authority over governance. It is another to make it happen. That is not to denigrate the state's legitimate authority. It is merely to affirm that under conditions where the state's governing capacity is problematic – something which is apparent at both domestic and transnational levels – one should explore a variety of auspices through which those same desired normative ends could be pursued (Johnston 2006b: 50, italics in original)

The above examples from different regulatory arenas have shown that non-state nodes, including those that are iconic of a new global civil society, possess or have the potential to possess the requisite knowledge, capacities and resources to monitor, and even to create, normative standards that guide them in their mix of governance functions. The virtue of the global civil society perspective, which we see in action with human rights NGOs, is its emphasis on local actors and their situated knowledge of regulatory issues that can be fed into the regulatory activities of transnational nodes and networks.

In simple terms, what seems to matter in the design of optimal regimes for the governance of governance is the right mix of 'upwards', 'downwards' and 'horizontal' processes (Scott 2000) that links up the activities of state and non-state nodes in ways that compensate for the weaknesses of each process on its own (see Goodin 2003). The involvement of global civil society in a range of governance arenas reflects an integration of top-down, bottom-up and horizontal coordination. If we return to the realm of policing, concrete proposals have been made in relation to the bolstering of *local* regulatory structures, drawing from the knowledge and capacities of stakeholder groupings in the strategic and financial steering of public *and* private service providers. The significance of these proposals is that they have accepted the reality of nodal security provision.

One example comes from Loader, who advances the idea of national, local and regional 'policing commissions' (Loader 2000; see also Jones 2003b: 618–20). Such commissions would engage in both procedural and outcome-based regulation of security based on three key principles. The first principle, 'a politics of recognition', requires commissions to take into account the view of all individuals and groups likely to be affected by particular forms of security provision in spaces to which such individuals and groups would have access (including 'mass private property'). This deliberative requirement is designed to ensure that there is no capture of the security agenda by dominant groups, such as purchasers with buying power (Loader 2000: 337–8).

The second principle, 'a politics of human rights', would ensure that the strategic decisions of a commission would not '[prejudice] the active rights of any individual or social group affected by the decision', or '[act] in ways that are disproportionately detrimental to the other interests and aspirations of such individuals and groups' (2000: 338).

The third principle, 'a politics of allocation', would require commissions to ensure the equitable distribution of security resources, both by monitoring, for example, the 'over-policing' of certain populations or the under-provision to those without buying power (2000: 338).

Based on these principles the policing commissions would undertake three main functions. First, the role of 'policy formation/ strategic coordination' would entail the development of overall policies, including the establishment of '*policing* plans within which *the police* are likely to play a significant but by no means the only part' (2000: 339, italics in original). The second function, of 'authorizing, licensing, subsidizing', could include the operation of a licensing system for private security as well as the administration of tenders and contracts to providers in line with established policing plans. Loader also suggests that commissions 'make good inequities in the (local) provision of policing and security' which could involve purchasing additional services for disadvantaged communities (2000: 339).

Third, Loader imagines the commissions as meta-monitors, inspectors and evaluators. 'This aspect of the commission's work', he argues, 'would be concerned with bringing the constituent nodes of the policing network to democratic account' (2000: 339). Central to this is research and reflection on security issues and trends and also on the network as a whole. Commissions could demand reports from particular nodes within the network as well as monitor both progress towards outcomes stipulated in policing plans and compliance with democratic standards (2000: 340).

A very similar agenda is found in the report of the Independent Commission on Policing in Northern Ireland (Patten Commission) which suggested the establishment of a Polic*ing* Board that would have more authority than established police boards in governing various forms of policing (Patten 1999; Shearing 2000; Walker 2000; Jones 2003a; Kempa in press). A central plank of the Commission's recommendation is the idea that the public police should be directed and controlled as simply one institution among several that are engaged in the governance of security. To this end the Commission recommended that the Policing Board control a *security* rather than a police budget, implying that the police would be required to compete for contracts within a policing market.

This proposal endorses a contractual system for the provision of policing that emphasizes a market-based approach to police accountability. In developing this approach the Commission drew upon ideas from both the legal and the political accountability approaches while at the same time seeking to extend them by responding to longstanding criticisms of both these approaches. For instance, the Commission argued that the concept of 'operational independence' is not the best way of reconciling political direction with the space that police require to carry out their functions. In place of the concept of operational independence they propose the concept of 'operational responsibility', which encapsulates the idea that while police should be responsible for carrying out directions and should have the freedom to do so, this does not mean that they should be able to operate outside of a framework of democratic scrutiny. Accordingly, the Commission proposed that the Policing Board have far-reaching powers to scrutinize what police do after the fact and argued that while police should be able to make decisions about how to implement policy and abide by the law, their operation practices should be open to scrutiny.

To realize this scrutiny the Commission endorsed a very robust complaints process and a very far-reaching audit process. The Commission's proposals provide that the Policing Board not only be able to audit police practice themselves but that they should be able to complement their own resources with resources commercially available. In addition, they proposed the establishment of an independent Oversight Commissioner who would be required to report publicly on the extent to which the Policing Board and the police were living up to their responsibilities.

If we refer back to the kinds of innovative regulatory developments we find beyond the realm of policing, there is clearly potential to think more innovatively about incorporating new mentalities, and the mechanisms that express them, as well as new nodal assemblages, with varying forms of knowledge and capacity, that can be linked together to enhance robustness in the governance of governance. There is no reason why this vision for enhanced hybrid governance cannot be explored in the transnational realm of (state *and* human) security provision. This normative agenda is clearly daunting because, apart from research conducted by scholars like Avant (2005), Singer (2001, 2003) and Johnston (2006b) there is still much we don't know about the range of formal and informal aspects of, for instance, the governance of commercial military service provision. While we know enough to have some rather big worries, it remains necessary to undertake a kind of explanatory 'mapping' of the governance regime as it currently exists (Wood 2006a). This would provide a sound empirical basis for then assessing regulatory failures, weaknesses and limitations of the system as a whole. Further, the question of how to address such weaknesses must itself be addressed through a deliberative process consisting of stakeholder groupings that are affected by, in one way or another, bad governance processes and outcomes (see Watchirs 2003 on the importance of this). If we draw lessons from fields outside of policing and (traditional) security, we know that robust civil society participation is essential. In short, it is not so much that state-based, command and control governance has failed in any particular setting. It's rather a matter of getting the mix right.

Conclusion

Drahos writes:

> Theories recognizing that regulation is more than a two-actor play and making a virtue of regulatory innovation are more likely to be able to provide strategies for dealing with problems relating to the supply and maintenance of public goods. In relation to global public goods where there is no sovereign provider, but simply a lot of imperfect multilateral institutions, considerable innovation is needed. (Drahos 2004: 323–4)

Existing approaches to the governance of governance already reveal considerable hybridity in the mechanisms and authorities that operate to shape the conduct of regulated actors towards sets of process- and outcome-level standards. The governance of the public police, where mentalities from business and the state sector are melded, provides one snapshot of a much larger nodal reality.

Our position is that much more can be done in exploring innovative combinations of mechanisms and authorities that can be assembled and structurally coordinated to the end of deepening commitment on the part of security providers (both state and non-state) to the normative standards that characterize our increasingly global world. The standards being established and promoted by nodes in global civil society rest rather squarely within a broader 'human security' orientation. Perhaps one of the biggest challenges in crafting governance designs for the future will be the establishment of appropriately deliberative structures and processes that allow for competing normative objectives to be articulated, weighed and aligned and competing mechanisms to be given equal consideration. Before this design phase begins, however, we must get our 'maps' right of what currently exists, central to which is the acceptance that governance is indeed more than a two-actor play.

Note

1 We are incredibly grateful to John Braithwaite for his insights on the 'governance of governance'.

Conclusion

In the preceding chapters we have attempted to set out the beginnings of a vision for the democratic and effective governance of security within a nodal world. In taking up this challenge we have been participating, and continue to participate, in a vibrant dialogue about governance within a 'polycentric' (McGinnis 1999a) and global world. Our objective, like that of Peter Drahos (Drahos and Braithwaite 2002; Drahos 2004, 2005a, 2005b), John Braithwaite (Braithwaite and Drahos 2000; Braithwaite 2002, 2004, 2006), Colin Scott (2000, 2002, 2004) and others from whom we have drawn inspiration, is to understand and re-imagine nodal governance in ways that recognize and promote established democratic values.

As part of this dialogue we have articulated a conception of the 'governance of security' which we have been developing with others over a period of several years (Shearing 1996; Shearing and Wood 2000; Johnston and Shearing 2003; Wood and Dupont 2006). In so doing we have sought to combine sociological thinking with a conception of governance that draws from Foucault's notion of 'government' (Foucault 1991) and Latour's understanding of 'power' (Latour 1986). These analytical strands support the view that there are different modes of making up the world that make possible different ways of governing it. Each of these modes may be thought of, following Foucault, as constituting a governing mentality. Each 'governmentality', to use Foucault's (1991) term, creates a particular set of possibilities for governing. The top-down, force-focused way of governing that we have associated with Hobbes constitutes one way, but only one way, of making up the world and acting upon it.

Foucault saw this Hobbesian conception of governance as essentially negative (Foucault 1990). He did so because it is built around the notion of constraint; it imagines governance as constraining rather than as producing action. This sovereign-based conception has dominated and continues to dominate the way we think about and engage in governance. Nevertheless, it is a conception that is unable to grasp the way in which governance has been developing because it sees power as owned, and centrally located, when in fact it is now increasingly dispersed and embedded in many different locations. It is everywhere, Foucault argues, not because it embraces and touches everything but because 'it comes from everywhere' (Foucault 1990).

This is an understanding that Foucault and a number of contemporary scholars who have been influenced by his thought have argued has come to be increasingly understood and acted upon by those who have sought to govern. This has led to the emergence of new ways of making up the world as governable. A new governmentality, Foucault argued, has been emerging that has shaped and is shaping the flow of events in the world. These emerging ideas have meant that governance has been made up and practised in ways that are different from how it is made up and practised within a Hobbesian framework. This has affected the way the police have developed and are developing. It has also shaped the environment in which they are governed by others. This has not meant that the Hobbesian mode of governance has disappeared, but rather that it has now become imbricated with other ways of making up and practising governance. These different dreams of governance coexist and meld in different ways. In the pages that follow we review the main explanatory and normative strands of this book.

Explanatory themes

The goal of this book has been to explicate a 'nodal' conception of security governance, one that accentuates diversity and hybridity in governing mentalities, institutions and practices. At an explanatory level, we argue that the nodal governance approach provides a useful analytical guide in mapping this diversity and hybridity. Normatively, we suggest that this analytics can inform innovations in the governance of security designed to alter existing nodes and nodal relationships and to foster new mentalities and construct new organizational assemblages.

The nodal perspective draws from previous work by Bayley and Shearing on the 'multilateralization' of policing (Bayley and Shearing 1996, 2001), an idea that seeks to distinguish 'auspices' from 'providers' of governance. This idea is based on the assumption that in order to understand what is happening to contemporary governance it is necessary to deploy an analytic framework that gives no *analytic* (as distinct from empirical) priority to states as either auspices or providers of governance. The nodal perspective, which seeks to build upon this idea, leaves open, both at an empirical and a normative level, what the position of states within governance was, or should be, at any space-time moment. It imagines governance as contested and sees states as comprising historically important sites of governance that increasingly exist within the contexts of other governmental nodes.

Governing nodes (see Burris 2004) are organizational sites (institutional settings that bring together and harness ways of thinking and acting) where attempts are made to intentionally shape the flow of events. Nodes govern under a variety of circumstances, operate in a variety of ways, are subject to a variety of different constraints, are motivated by a variety of objectives and concerns, and engage in a variety of different actions to shape the flow of events. Nodes relate to one another, and attempt to mobilize and resist one another, in a variety of ways so as to shape matters in ways that promote their objectives and concerns. Nodal governance is diverse and complex.

The world of nodal governance is hierarchically structured. Hierarchy emerges as auspices of governance seek to align the activities of other nodes in support of their governance objectives. Many hierarchies coexist. There are multiple 'tops' and many 'downs' and they overlap. Auspices often contest one another so that nodal governance is by no means a smooth coordinated business, although much coordination does take place. Coordination typically takes place when nodes successfully enrol other nodes in their agendas to shape the flow of events.

Governments govern through a variety of nodes. They are themselves also enrolled by nodes that govern through them and their agencies. Sometimes these other nodes are enrolled and recruited as part of rule 'at a distance' (Latour 1987; Rose and Miller 1992) arrangements that state nodes establish to realize their governing agendas. This certainly happens but it is not all that happens. Not all nodes form part of state rule-at-a-distance hierarchies. Nodes beyond the state – often tacitly or explicitly supported by states – engage other nodes, including state nodes, as part of their rule-at-a-distance strategies. These hierarchies intersect in different ways with state-based hierarchies. We have argued that it is not simply the case that the activities of corporate nodes are coordinated by states so as to meet the governing objectives of state nodes. While this does happen, to see this as the exclusive picture is to engage in an error of presenting one part of this picture as the whole.

In understanding how nodes and nodal arrangements have been transformed across different times and spaces we argued that it is useful to focus attention on subtle shifts in ways of thinking that prompt new ways of acting. There are periods when transformations in the way the world is governed take place within established mentalities. During these periods change is a matter of fine-tuning. It is incremental. Every so often, however, change becomes more radical. This happens when new ways of thinking are invoked, new institutional forms are developed and new technologies are deployed. Toffler has used the metaphor of 'waves' to understand these broader, paradigmatic shifts (1980). This imagery suggests that when broader shifts occur they draw upon what has gone before, in the same way as a wave in the ocean draws water into it as it is formed.

This metaphor cautions us, when we are looking at even very radical change, to be aware of and pay attention to continuities. We attempted to illustrate this in our analysis of waves in public policing. Innovation should be conceived of as novelty as well as reconfiguration. As with an ocean, much has already taken place before a governance wave rises out of a swell into a shape that is explicitly recognized as new. In his study of transformations in the legal system of Mozambique, de Sousa Santos provides a similar analysis, observing that 'ruptures coexisted with continuities, blending explicit and self-proclaimed ruptures with unspoken continuities and so giving rise to very complex legal and institutional constellations and hybridizations' (de Sousa Santos 2006: 48).

The transformations that have been taking place globally illustrate the extent to which hybridity in the mentalities, strategies and institutional arrangements of governance has been accomplished. We discussed how the governance of security has come to pervade other domains of governance – and vice versa – through a process of 'securitization' (Buzan *et al.* 1998). Particularly important has been the development of the concept of 'human security'. This concept seeks to focus governance on the well-being of individuals across a range of domains. In particular, it seeks to extend the objective of security governance – understood traditionally as promoting states where people are free from threats to their bodies and property (what

Thomas Hobbes (1651) thought of as 'peace') – to one of freedom from a much wider array of threats. The Hobbesian conception of peace has long been part of our understanding of policing as the business of 'keeping the peace'. With the emergence of the notion of human security, peacekeeping and the more proactive term 'peacebuilding' has taken on a much more extensive ambit. The idea of human survival that is inherent in the idea of keeping the peace is now given an even wider purchase.

Together the nodal nature of governance and the expansion of the meanings we give to 'security' have served to constitute the governance of security in very expansive ways. This, as we have seen, creates analytic challenges in developing conceptual frameworks that are sufficiently general to ensure that while they include and pay considerable attention to state nodes – and the traditional imaginings of security that shape their mentalities and practices – they are not state-focused. Once we have such frameworks they must then be used to provide descriptive and explanatory accounts that capture the nature of contemporary governance. Much of this book has been about attempting to do just this.

Normative themes

Our account of nodal governance explores both the reality of what has occurred and constructs a normative theory of what possibilities might exist for governing democratically within a nodal world in ways that ensure that common goods will continue to be provided to a whole variety of different constituencies. Within this framework the possibility and desirability of drawing on the capacities of states as sources of regulation (the governance of governance) as well as provision is regarded as one of a range of possibilities to be considered in delivering common goods. In discussing possibilities for the future we argued that governance must take place within complex assemblages and networks of direction and provision if effective democratic governance is to flourish as we move into the twenty-first century. In building new ideas we need to draw inspiration from the achievements of governance within a more state-centred world without being overwhelmed by the thinking and institutions that dominated the nineteenth and the twentieth centuries.

In making this conceptual shift to nodal governance the debate over the provision of public goods focuses not simply on the question of how to construct an overarching public interest that includes very wide collectivities (that could conceivably extend to the planet as a whole and to multiple species). It also considers the question of how valued collective goods are to be identified and how they are to be produced and distributed within a nodal context. What is needed is a conception of complex assemblages of auspices and providers that include 'the hundreds of thousands of groups that make up international civil society' (Drahos 2004: 336). As we discussed in Chapter 5, this understanding of governance as decentred and multinodal promotes arrangements whereby 'state and non-state actors regulate each other's capacities to provide, access, and distribute public goods' (2004: 321).

Democratic governance does not guarantee that the core values that many humans now widely believe are keys to successful life on this planet will be realized, or that human life (and the life of many other species whose existence is affected by human practices) will flourish on this planet. The claim that effective democratic governance will enable life to flourish rests on the principle, or more accurately the hope, that human beings collectively have the wisdom and the capacity to make good governance decisions. One of the most telling empirical studies of this principle and its importance for the most significant human challenge of our time – namely, the sustainability of human life on this increasingly fragile planet – is the analysis provided by Diamond (2005) of the conditions under which human life has collapsed and then sustained over time. He makes abundantly clear that the continuity of human life on this planet has never been something that could be taken for granted and that human societies have collapsed, and by implication will continue to do so, to the extent to which they do not govern well. In our new globally integrated world, such collapses are unlikely to be ones that will affect some collectivities but not others. If we have another major societal collapse it is likely to be one with global consequences for ourselves and for other species. One of Diamond's most telling conclusions is that one of the conditions for collapse is a divided society in which the strong both make the decisions and can, for a while, externalize the negative consequences of these decisions so that others bear the burden. Eventually, the burdens can no longer be externalized and the society collapses.

In our contemporary world of nodal governance there are deep social fissures exacerbated by wealth and power which have taken on new contours in the context of globalization. As we have discussed, the 'new wars' are an expression of this harsh reality. Linked to such fissures are deficits in the capacity of actors to govern in ways that promote their collective interests. In recognizing this we accept that our world will continue to be governed nodally and that the actors who do have the capacity to shape the contours of governance to promote their objectives will do so across a whole series of nodes and networks. This, we believe, is the fact of the matter now and will remain so for some time to come. Those who benefit from nodal governance are now, and will continue, doing so. The reality is that there are 'weak' actors and 'strong' actors in the game of nodal governance (Braithwaite 2004; Braithwaite and Drahos 2000; Drahos and Braithwaite 2002).

In responding to this profound normative concern we are taking a stance that sets us apart from the mainstream. Scholars located within this stream seek to respond to the deficits we have identified by arguing for solutions that seek in one way or another to shift governance in a more monocentric direction, either by advocating strategies that would strengthen states or by seeking to strengthen supra-state developments. Our reasons for swimming outside of the mainstream are pragmatic and strategic. We suspect that whatever happens to strengthen and democratize states and supra-states will reshape our nodal world, but it will not make it less nodal.

Drawing from the insights of Braithwaite and Drahos (2000), we have argued that it is essential to take stock of, and to develop, strategies for shaping nodal governance in ways that will benefit disadvantaged populations. This requires a 'democratic experimentalism' (Dorf and Sabel 1998). Braithwaite (2004), in seeking to promote this experimentalism, has taken a significant step. He identifies strategies that the winners in the global game of nodal governance have used to advantage themselves. The perspective he takes is to ask how strategies of the strong might be reshaped as strategies for the weak that can be used by them to strengthen their ability to play the nodal governance game. Underlying this stance is the implicit claim that it is not until these weak players become stronger that nodal governance will become more equitable in its outcomes and more democratic in its processes.

We have taken a very similar position in the normative research we have undertaken within poor communities in both Argentina and South Africa (Shearing and Wood 2000, 2003b; Shearing 2001b; Wood and Font in press). Here we asked how poor communities can best take advantage of nodal governance to govern security in ways that are more effective for them, provide them with greater self-direction in setting security agendas, promote democratic and human rights values, and relate to state agencies in ways that promote effective and acceptable state policing. The assumption underlying this form of experimentalism is that it is the weak who must ultimately be the ones who decide what will constitute, and best provide for, their well-being. To the extent that they are able to promote greater well-being for themselves, and to the extent that their hand in contests over governance is strengthened, we will live in a more democratic world. This will be a world in which the many – to whom established democratic mechanisms, as a matter of fact, pay very little attention and indeed have largely disenfranchised – will become more self-directed.

Of course, such an agenda for enhancing the power of the weak does not seek to dispel every worry we have about nodal governance. The capacity of weak actors to play greater roles as auspices of governance does not preclude the possibility that such auspices, or any others for that matter, will make decisions or engage in practices that run counter to the core values we have identified. As we discussed in Chapter 5, there are definite concerns about this in relation to the decisions and practices of commercial auspices and providers of security, and clearly there has always been the potential for, and reality of, police institutions themselves breaching the very laws they were established to protect. Indeed within states it is becoming increasingly difficult to argue that democratic governments express the general will of their citizenry when so many opportunities exist for special interests to capture governments for their own purposes. From a broader human security perspective, we know that states (including states that are seen as democratic exemplars) can potentially do more harm than good. And even if they do not harm everyday people in a direct manner, they can allow the objectives and concerns of their own citizens to dominate globally at the expense of others. We are thus faced with the challenge of 'governing governance' in a nodal world.

If we accept that we live in a polycentric world with multiple sources of governance organized across a range of different governance hierarchies, then to think of exercising normative direction over governance through states and supra-state agencies alone is to deliberately avoid facing up to the normative challenges that face us. To say this is not to suggest that accountability to and through states (and supra-states) involving the deployment and enhancement of the various mechanisms developed to do this is not sensible. As we have argued, it clearly is. It is also not to suggest that the values and objectives that these mechanisms promote and seek to realize should not be central to developing normative frameworks for constraining and directing nodal governance. What is required is not an either/or choice between market-oriented and state-dominated approaches or between state and non-state conceptions of governance, but rather an approach that eschews the weaknesses of each and at the same time identifies and mobilizes other possibilities.

In Chapter 5 we explored the theme of hybridity in governing mechanisms and in the types of governing authorities that have developed in order to shape the conduct of state and non-state nodes. In our analysis we drew broadly from developments in the governance of governance in fields beyond the traditional domains of policing and security. We argued that it is in this broader field – where we find innovations in areas such as environmental and agricultural regulation – that we can draw much inspiration in designing new models for the (nodal) governance of (nodal) governance. Indeed, it is from a human security standpoint that one discovers the connections between the governance of security governance and the governance of so many other areas of life that have a direct impact on human well-being and flourishing on this planet.

The assumption (and hope) behind such a quest to re-imagine and reinvent the governance of governance is that human beings will be able to deliver collective goods more wisely and in ways that not only benefit the strong, but will also benefit most of humankind. Our belief is that ultimately we have no alternative but to rely on the goodwill and intelligence of human beings. What this goodwill and intelligence produces must be constantly reinvented as collective life evolves.

In conclusion, perhaps the most fundamental thought underlying the idea of democracy is that in the final analysis it is not the wisdom of a Leviathan that we must rely upon, be this secular or non-secular, but the collective wisdom of human beings to promote conditions for human well-being and, more broadly, planetary well-being. The question we face today is: how are we, as human collectivities operating in a nodal world of governance, to rely on our wisdom to promote our security? Just how wisdom is produced and tapped is different in different locations and involves voter preferences, deliberative forums of various kinds, knowledge and expertise and so on. Now more than ever, deliberative mechanisms that centre on the knowledge and capacity of weak actors in building new ideas for security governance appear to constitute an essential step in addressing the governance deficits of our nodal world. For Kaldor, what is required is a 'conversation in which we talk about our moral concerns, our passions, as well as thinking through the best way to solve problems. A conversation in which the participants are not just those who can travel and communicate across long distances, but also ordinary men, women and children' (Kaldor 2003: 160). The reasoning underlying such a position is not that we will prove to be wise enough, but that we have no other option. We are all that we have. We cannot look outside of or beyond ourselves for wisdom. We are it.

Bibliography

Angel-Ajani, A. (2003) 'A Question of Dangerous Races?', *Punishment and Society*, 5(4): 433–48.

Arquilla, J. (ed.) (2001) *Networks and Netwars: The Future of Terror, Crime, and Militancy*, Santa Monica, CA: RAND.

Audit Commission (1993) *Helping With Enquiries: Tackling Crime Effectively*, London: Audit Commission.

Auditor General of Canada (2002) *Report of the Auditor General of Canada to the House of Commons*, Ottawa: Office of the Auditor General of Canada.

Auditor-General's Office, ACT (2004) *The Administration of Policing Services. Performance Audit Report*, Canberra: Auditor-General's Office.

AusAID (2004) *Food Security Strategy*, Canberra: Commonwealth of Australia.

Avant, D. (2005) *The Market for Force: The Consequences of Privatizing Security*, Cambridge: Cambridge University Press.

Ayling, J. and Grabosky, P. (2006) 'Policing by Command: Enhancing Law Enforcement Capacity through Coercion', *Law and Policy*, 28(4).

Ayling, J., Grabosky, P. and Shearing, C. (2006) 'Harnessing Resources for Networked Policing', in J. Fleming and J. Wood (eds) *Fighting Crime Together: The Challenges of Policing and Security Networks*, Sydney: University of New South Wales Press.

Ayres, I. and Braithwaite, J. (1992) *Responsive Regulation: Transcending the Deregulation Debate*, New York, NY: Oxford University Press.

Barker-McCardle, J. (ed.) (2000) *Intelligence-led Policing, Crime Reduction, Performance Management, Regulators, Rat-catchers and Thrusters. Organizational Performance Measurement*. Aylmer, Ont.: Ontario Police College.

Barry, B. (1999) 'Statism and Nationalism: A Cosmopolitan Critique', in I. Shapiro and L. Brilmayer (eds) *Global Justice*, New York, NY: New York University Press.

Bayley, D. (1983) 'Accountability and Control of Police: Some Lessons for Britain', in T. Bennett (ed.) *The Future of Policing*, Cambridge: Institute of Criminology, pp. 146–62.

Bayley, D. (1988) 'Community Policing: A Report from the Devil's Advocate', in J. R. Greene and S. D. Mastrofski (eds) *Community Policing: Rhetoric or Reality?*, New York: Praeger.

Bayley, D. (1995) 'Getting Serious about Police Brutality', in P. Stenning (ed.) *Accountability for Criminal Justice: Selected Essays*, Toronto: University of Toronto Press, pp. 93–109.

Bayley, D. (1999) 'Remarks on "Zero Tolerance" at SUNY Albany Conference on Zero Tolerance Policing Initiatives – Legal and Policy Perspectives', *Criminal Law Bulletin*, 35(4): 369–70.

Bayley, D. and Shearing, C. (1996) 'The Future of Policing', *Law and Society Review*, 30(3): 585–606.

Bayley, D. and Shearing, C. (2001) *The New Structure of Policing: Description, Conceptualization, and Research Agenda*, Washington, DC: National Institute of Justice.

Beccaria, C. (1764) *On Crimes and Punishments* (trans. H. Paolucci), Indianapolis, IN: Bobbs-Merrill (reprinted 1963).

Beare, M. E. (ed.) (2003) *Critical Reflections on Transnational Organized Crime, Money Laundering, and Corruption*, Toronto: University of Toronto Press.

Bernasek, A. (2003) 'Banking on Social Change: Grameen Bank Lending to Women', *International Journal of Politics, Culture and Society*, 16(3): 369–85.

Bernstein, S. and Cashore, B. (2005) *The Two-Level Logic of Non-State Market Driven Global Governance*, Toronto: Department of Political Science, University of Toronto.

Bigo, D. (2000) 'Liaison officers in Europe: New Officers in the European Field', in J.W.E. Sheptycki (ed.) *Issues in Transnational Policing*, London: Routledge, pp. 67–99.

Bittner, E. (1970) *The Functions of Police in Modern Society: A Review of Background Factors, Current Practices, and Possible Role Models*, Chevy Chase, MD: National Institute of Mental Health.

Bittner, E. (1990) *Aspects of Police Work*, Boston, MA: Northeastern University Press.

Black, J. (2000) 'Decentring Regulation: Understanding the Role of Regulation and Self-regulation in a "Post-regulatory" World', *Current Legal Problems*, 54: 103–146.

Blair, I. (1998) 'Where Do Police Fit into Policing?', paper presented at *Annual Conference of the Association of Chief Police Officers*, London, July.

Blair, I. (2003) 'Surprise News: Policing Works – A New Model of Patrol', paper presented at *In Search of Security: An International Conference on Policing and Security*, Montreal, Law Commission of Canada, February.

Blair, I. (2005) *Dimbleby Lecture*, available online at http://news.bbc.co.uk/1/hi/uk/4443386.stm /(accessed 10 June 2006).

Blair, I. (2006) 'Surprise News: Policing Works', *Police Practice and Research*.

Bourdieu, P. (1986) 'The Forms of Capital', in J. G. Richardson (ed.) *Handbook of Theory and Research for the Sociology of Education*, New York, NY: Greenwood Press, pp. 241–58.

Bowling, B. (1999) 'The Rise and Fall of New York Murder: Zero Tolerance or Crack's Decline?', *British Journal of Criminology*, 39(4): 531–54.

Bradley, D. (1994) 'Problem-Oriented Policing: Old Hat or New Thing?', paper presented at *Critical Issues Seminar on Problem Oriented Policing*, Manly: Australian Police Staff College.

Braithwaite, J. (1985) *To Punish or Persuade: Enforcement of Coal Mine Safety*, Albany, NY: State University of New York Press.

Braithwaite, J. (1989) *Crime, Shame and Reintegration*, Cambridge: Cambridge University Press.

Braithwaite, J. (1993) 'Responsive Regulation for Australia', in P. Grabosky and J. Braithwaite (eds) *Business Regulation and Australia's Future*. Canberra: Australian Institute of Criminology.

Braithwaite, J. (2000a) 'The New Regulatory State and the Transformation of Criminology', *British Journal of Criminology*, 40: 222–38.

Braithwaite, J. (2000b) *Regulation, Crime, Freedom*, Dartmouth: Ashgate.

Braithwaite, J. (2002) *Restorative Justice and Responsive Regulation*, Oxford: Oxford University Press.

Braithwaite, J. (2003a) 'What's Wrong with the Sociology of Punishment?', *Theoretical Criminology*, 7(1): 5–28.

Braithwaite, J. (2003b) 'Restorative Justice and Corporate Regulation', in E. Weitekamp and H.-J. Kerner (eds) *Restorative Justice in Context: International Practice and Directions*, Cullompton: Willan Publishing.

Braithwaite, J. (2004) 'Methods of Power for Development: Weapons of the Weak, Weapons of the Strong', *Michigan Journal of International Law*, 26(1): 297–330.

Braithwaite, J. (2006) 'Peacemaking Networks and Restorative Justice', in J. Fleming and J. Wood (eds) *Fighting Crime Together: The Challenges of Policing and Security Network*, Sydney: University of New South Wales Press.

Braithwaite, J. and Drahos, P. (2000) *Global Business Regulation*, Cambridge: Cambridge University Press.

Braithwaite, J. and Pettit, P. (1990) *Not Just Deserts: A Republican Theory of Criminal Justice*, Oxford: Clarendon Press.

Bratton, W. (1998) 'Crime is Down in New York City: Blame the Police', in N. Dennis (ed.) *Zero Tolerance: Policing a Free Society*, London: IEA Health and Welfare Unit.

Bratton, W. and Knobler, P. (1998) *Turnaround: How America's Top Cop Reversed the Crime Epidemic*, New York, NY: Random House.

Brown, M. (1981) *Working the Street: Police Discretion and the Dilemmas of Reform*, New York, NY: Russell Sage Foundation.

Burris, S. (2004) 'Governance, Microgovernance and Health', *Temple Law Review*, 77: 335–59.
Burris, S. (2006) 'From Security to Health', in J. Wood and B. Dupont (eds) *Democracy, Society and the Governance of Security*, Cambridge: Cambridge University Press, pp. 196–216.
Burris, S., Drahos, P. and Shearing, C. (2005) 'Nodal Governance', *Australian Journal of Legal Philosophy*, 30: 30–58.
Buzan, B., Wæver, O. and de Wilde, J. (1998) *Security: A New Framework for Analysis*, Boulder, CO: Lynne Rienner.

Canadian Association of Chiefs of Police (2000) *A National Use of Force Framework*, available online at www.cacp.ca/(accessed 10 June 2006).
Cartwright, J. (2004) 'From Aquinas to Zwelethemba: A Brief History of Hope', *Annals of the American Academy of Political and Social Science*, 592 (March): 166–84.
Cashore, B. (2002) 'Legitimacy and the Privatization of Environmental Governance: How Non-State Market-Driven (NSMD) Governance Systems Gain Rule-Making Authority', *Governance: An International Journal of Policy, Administration, and Institutions* 15(4): 503–29.
Cashore, B., Auld, G. and Newsom, D. (2004) *Governing Through Markets: Forest Certification and the Emergence of Non-State Authority*, New Haven, CT: Yale University Press.
Castells, M. (2000) *The Rise of the Network Society: The Information Age; Economy, Society and Culture*, Oxford: Blackwell.
Centro de Estudios Legales y Sociales (CELS) and in conjunction with a network of organizations (2004) *Más derechos, más seguridad, más seguridad, más derechos: Políticas públicas y seguridad en una sociedad democrática*, Buenos Aires: CELS and partner organizations.
Cerven, J. and Ghazanfar, S.M. (1999) 'Third World Microfinance: Challenges of Growth and Possibilities for Adaptation' *Journal of Social, Political, and Economic Studies*, 24(4): 445–62.
Chan, J. (1996) 'Changing Police Culture', *British Journal of Criminology*, 36(1): 109–34.
Chan, J. (1997) *Changing Police Culture: Policing in a Multicultural Society*, Melbourne: Cambridge University Press.
Chan, J. (2001a) 'Negotiating the Field: New Observations on the Making of Police Officers', *Australian and New Zealand Journal of Criminology*, 34(2): 114–33.
Chan, J. (2001b) 'The Technological Game: How Information Technology is Transforming Police Practice', *Criminal Justice*, 1(2): 139–59.
Chan, J., Devery, C. and Doran, S. (2003) *Fair Cop: Learning the Art of Policing*, Toronto: University of Toronto Press.
Cherney, A., O'Reilly, J. and Grabosky, P. (2005) *The Governance of Illicit Synthetic Drugs*. Monograph Series No. 9. Canberra: National Drug Law Enforcement Research Fund.

Cherney, A., O'Reilly, J. and Grabosky, P. (2006) 'The Multilateralization of Policing: The Case of Illicit Synthetic Drug Control', *Police Practice and Research*, 7(3): 175–6.

Christie, N. (1978) 'Conflicts as Property' *British Journal of Criminology,* 17: 1–15.

Cohen, S. (1995) 'State Crimes of Previous Regimes: Knowledge, Accountability and the Policing of the Past', *Law and Social Inquiry,* 20(1): 7–50.

Commission on Human Security (2003) *Final Report*, New York, NY: United Nations Commission on Human Security.

Cooley, D. (ed.) (2005) *Re-imagining Policing in Canada,* Toronto: University of Toronto Press.

Cope, N. (2003) 'Crime Analysis: Principles and Practice', in T. Newburn (ed.) *Handbook of Policing*, Cullompton: Willan Publishing.

Cope, N. (2004) 'Intelligence Led Policing or Policing Led Intelligence?', *British Journal of Criminology,* 44: 188–203.

Courville, S. (2003a) 'Social Accountability Audits: Challenging or Defending Democratic Governance?', *Law and Policy* 25(3): 269–97.

Courville, S. (2006) 'Understanding NGO-Based Social and Environmental Regulatory Systems: Why We Need New Models of Accountability', in M.W. Dowdle (ed) *Public Accountability: Designs, Dilemmas and Experiences,* Cambridge: University of Cambridge Press, pp. 271–300.

Crawford, A. (1997) *The Local Governance of Crime: Appeals to Community and Partnerships*, Oxford: Clarendon Press.

Crawford, A. (1998) *Crime Prevention and Community Safety: Politics, Policies and Practices*, London: Longman.

Crawford, A. (2003) '"Contractual Governance" of Deviant Behaviour', *Journal of Law and Society,* 30(4): 479–505.

Crawford, A. (2006) 'Policing and Security as "Club Goods": The New Enclosures?', in J. Wood and B. Dupont (eds) *Democracy, Society and the Governance of Security*, Cambridge: University of Cambridge Press, pp. 111–138.

Crawford, A. and Lister, S. (2004) 'The Patchwork Shape of Reassurance Policing in England and Wales: Integrated Local Security Quilts or Frayed, Fragmented and Fragile Tangled Webs?', *Policing: An International Journal of Police Strategies and Management*, 27(3): 413–30.

Crawford, A., Lister, S., Blackburn, S. and Burnett, J. 92005) *Plural Policing: The Mixed Economy of Visible Patrols in England and Wales.* Bristol: The Policy Press.

Cuneen, C. (1999) 'Zero Tolerance Policing and the Experience of New York City', *Current Issues in Criminal Justice,* 10(3): 299–313.

Daley-Harris, S. (2003) *State of the Microcredit Summit Campaign Report 2003*, available online at http://www.microcreditsummit.org/pubs/reports/socr/(accessed 10 June 2006).

Daniels, R. J., Macklem, P. and Roach, K. (eds) (2001) *The Security of Freedom: Essays on Canada's Anti-Terrorism Bill*, Toronto: University of Toronto Press.

Datta, R. (2003) 'From Development to Empowerment: The Self-Employed Women's Association in India', *International Journal of Politics, Culture and Society*, 16(3): 351–68.

De Lint, W. (1997) 'The Constable Generalist as Exemplary Citizen, Networker and Problem-Solver: Some Implications', *Policing and Society*, 6: 247–64.

De Lint, W. (1998) 'New Managerialism and Canadian Police Training Reform', *Social and Legal Studies*, 7(2): 261–85.

Dennis, N. and Mallon, R. (1997) 'Confident Policing in Hartlepool', in N. Dennis (ed.) *Zero Tolerance*, London: Institute of Economic Affairs.

Department of Premier and Cabinet (2001) *Growing Victoria Together*, Melbourne: Department of Premier and Cabinet Victoria.

De Sousa Santos, B. (2006) 'The Heterogeneous State and Legal Pluralism in Mozambique', *Law and Society Review*, 40(1): 39–75.

Deukmedjian, J. (2002) *The Evolution and Alignment of RCMP Conflict Management and Organizational Surveillance*, Toronto: Centre of Criminology, University of Toronto.

Diamond, J.M. (2005) *Collapse: How Societies Choose to Fail or Succeed*, New York, NY: Viking.

Dorf, M.C. and Sabel, C.F. (1998) 'A Constitution of Democratic Experimentalism', *Columbia Law Review*, 98: 267–473.

Drahos, P. (2004) 'The Regulation of Public Goods', *Journal of International Economic Law*, 7(2): 321–39.

Drahos, P. (2005a) 'Intellectual property and pharmaceutical markets: a nodal governance approach', *Temple Law Review*, 77: 401–24.

Drahos, P. (2005b) 'Winning Battles, Losing the War; Lessons for the Weak from the Negotiations over the Doha Declaration on TRIPS and Public Health', paper presented at *Trade Negotiation and Developing Countries: The Doha Round*, Brisbane.

Drahos, P. and Braithwaite, J. (2002) *Information Feudalism*, London: Earth Scan.

Drucker, P. (1994) 'The Age of Social Transformation', *The Atlantic Monthly*, November: 53–80.

Dryzek, J.S. (2000) 'Deliberative Democracy and Beyond: Liberals, Critics, Contestations', in W. Kymlicka, D. Miller and A. Ryan (eds) *Political Theory*, Oxford: Oxford University Press.

Duffield, M. (2001) *Global Governance and the New Wars*, London: Zed Books.

Duffield, M. (2002) 'War as a Network Enterprise: The New Security Terrain and its Implications', *Cultural Values*, 6: 153–65.

Duffield, M. (2005) 'Getting Savages to Fight Barbarians: Development, Security and the Colonial Present', *Conflict, Security and Development*, 5(2): 141–59.

Duffield, M. and Waddell, N. (2006) 'Securing Humans in a Dangerous World', *International Politics*, 43: 1–23.

Dupont, B. (2003) 'Public Entrepreneurs in the Field of Security: An Oral History of Australian Police Commissioners', paper presented at *In Search of Security: An International Conference on Policing and Security*, Montreal: Canada, Law Commission of Canada, February.

Dupont, B. (2006a) 'Power Struggles in the Field of Security: Implications for Democratic Transformation', in J. Wood and B. Dupont (eds) *Democracy, Society and the Governance of Security*, Cambridge: Cambridge University Press.

Dupont, B. (2006b) 'Mapping Security Networks: From Metaphorical Concept to Empirical Model', in J. Fleming and J. Wood (eds) *Fighting Crime Together: The Challenges of Policing and Security Networks*, Sydney: University of New South Wales Press.

Economist (2005) 'A Thicker Blue Line', 376: 41.

Einstadter, W. and Henry, S. (1995) *Criminological Theory: An Analysis of its Underlying Assumptions*, Fort Worth, TX: Harcourt, Brace and Co.

Ericson, R. (1981) 'Rules for Police Deviance', in C. Shearing (ed.) *Organizational Police Deviance*, Toronto: Butterworths.

Ericson, R. (1994) 'The Division of Expert Knowledge in Policing and Security', *British Journal of Sociology*, 45(2): 149–75.

Ericson, R. and Haggerty, K. (1997) *Policing the Risk Society*, Toronto: University of Toronto Press.

Espeland, W.N. (1998) 'Authority by the Numbers: Porter on Quantification, Discretion, and the Legitimation of Expertise', *Law and Social Inquiry*, 22(4): 1107–33.

Espeland, W.N. and Stevens, M.L. (1998) 'Commensuration as a Social Process', *Annual Review of Sociology*, 24: 313–43.

Fagan, J., Zimring, F.E. and Kim, J. (1998) 'Declining Homicide in New York City: A Tale of Two Trends', *Journal of Criminal Law and Criminology*, 88(4): 1277–1323.

Feinstein International Famine Centre (2004) *The Future of Humanitarian Action: Implications of Iraq and Other Recent Crises*, Boston, MA: Feinstein International Famine Centre/Friedman School of Nutritional Science Policy, Tufts University.

Fielding, N. and Innes, M. (2006) 'Reassurance Policing, Community Policing and Measuring Police Performance', *Policing and Society* 16(2): 127–45.

Fleming, J. and Wood, J. (2006) 'Introduction: New Ways of Doing Business, Networks of Policing and Security', in J. Fleming and J. Wood (eds)

Fighting Crime Together: The Challenges of Policing and Security Networks, Sydney: University of New South Wales Press.

Flood, B. (2004) 'Strategic Aspects of the UK National Intelligence Model', in J. Ratcliffe (ed.) *Strategic Thinking in Criminal Justice*, Sydney: Federation Press.

Foucault, M. (1977) *Discipline and Punish: The Birth of the Prison*, New York: Vintage Books.

Foucault, M. (1982) 'The Subject and Power', in H. L. Dreyfus and P. Rabinow (eds) *Michel Foucault: Beyond Structuralism and Hermeneutics*, Chicago, IL: University of Chicago Press.

Foucault, M. (1988) 'Power, Moral Values, and the Intellectual: An interview with Michel Foucault', *History of the Present*, Spring: 1–2, 11–13.

Foucault, M. (1990) *The History of Sexuality, Volume 1: An Introduction*, New York, NY: Vintage Books.

Foucault, M. (1991) 'Governmentality', in G. Burchell and C. Gordon (eds) *The Foucault Effect: Studies in Governmentality*, Chicago, IL: University of Chicago Press.

Foucault, M. (2003) *Society Must be Defended: Lectures at the College de France, 1975-76*, London: Allen Lane.

Fukuyama, F. (1992) *The End of History and the Last Man*, New York, NY: Free Press.

Fung, A. (2001) 'Accountable Autonomy: Toward Empowered Deliberation in Chicago Schools and Policing', *Politics and Society*, 29(1): 73–103.

Fung, A. (2004) *Empowered Participation: Reinventing Urban Democracy*, Oxford: Princeton University Press.

Garland, D. (1996) 'The Limits of the Sovereign State: Strategies of Crime Control in Contemporary Society', *British Journal of Criminology*, 40: 347–75.

Gilboy, J.A. (1998) 'Compelled Third-Party Participation in the Regulatory Process: Legal Duties, Culture, and Noncompliance', *Law and Policy*, 20(2): 135–55.

Gilling, D. (1997) *Crime Prevention: Theory, Policy and Politics*, London: UCL Press.

Goldsmith, A. (ed.) (1991a) *Complaints against the Police: The Trend to External Review*, Oxford: Clarendon Press.

Goldsmith, A. (1991b) 'External Review and Self-Regulation: Police Accountability and the Dialectic of Complaints Procedures', in A. Goldsmith (ed.) *Complaints against the Police: The Trend to External Review*, Oxford: Clarendon Press.

Goldsmith, A. (1995) 'Necessary but Not Sufficient: The Role of Public Complaints Procedures in Police Accountability', in P. Stenning (ed.) *Accountability for Criminal Justice*, Toronto: University of Toronto Press.

Goldsmith, A. (2000) 'An Impotent Conceit: Law, Culture and the Regulation of Police Violence', in T. Coady, S. James, S. Miller and M. O'Keefe (eds) *Violence and Police Culture*, Melbourne: Melbourne University Press.

Goldstein, H. (1979) 'Improving Policing: A Problem-Oriented Approach', *Crime and Delinquency*, 25 (April): 236–58 / reprinted (1991) in C.B. Klockars and S.D. Mastrofski (eds) *Thinking About Police: Contemporary Readings*, New York, NY: McGraw-Hill.

Goldstein, H. (1990) *Problem-Oriented Policing*, New York, NY: McGraw-Hill.

Goldstein, H. (1999) 'Remarks on "Zero Tolerance" at SUNY Albany Conference on Zero Tolerance Policing Initiatives – Legal and Policy Perspectives', *Criminal Law Bulletin*, 35(4): 370–73.

Goodin, R.E. (2003) 'Democratic Accountability: The Distinctiveness of the Third Sector', *European Journal of Sociology*, 44(3): 359–96.

Government of Ontario (1998) *Who Does What? Toward Implementation*, Toronto: Government of Ontario.

Grabosky, P. (1995) 'Using Non-Governmental Resources to Foster Regulatory Compliance', *Governance: An International Journal of Policy and Administration*, 8(4): 527–50.

Grabosky, P. and Braithwaite, J. (eds) (1993) *Business Regulation and Australia's Future*, Canberra: Australian Institute of Criminology.

Grameen Bank (2006) *Grameen Bank*, available online at http://www.grameen-info.org/(accessed 14 September 2006).

Granovetter, M.S. (1973) 'The Strengh of Weak Ties', *American Journal of Sociology*, 78(6): 1360–80.

Greene, J.A. (1999) 'Zero Tolerance: a Case Study of Police Policies and Practices in New York City', *Crime and Delinquency*, 45(2): 171–87.

Gross Stein, J. (2001) 'Network Wars', in R.J. Daniels, P. Macklem and K. Roach (eds) *The Security of Freedom. Essays on Canada's Anti-Terrorism Bill*, Toronto: University of Toronto Press.

Gunningham, N. and Grabosky, P. (1998) *Smart Regulation: Designing Environmental Policy*, Oxford: Clarendon Press.

Gunningham, N., Kagan, R.A. and Thornton, D. (2003) *Shades of Green: Business, Regulation and Environment*, Stanford, CA: Stanford University Press.

Gunningham, N., Kagan, R.A. and Thornton, D. (2004) 'Social License and Environmental Protection: Why Businesses Go Beyond Compliance', *Law and Social Inquiry*, 29(2): 307–41.

Haggerty, K. and Ericson, R. (1999) 'The Militarisation of Police in the Information Age', *Journal of Political and Military Sociology*, 27 (Winter): 233–55.

Hall, R.B. and Biersteker, T.J. (2002a) 'The Emergence of Private Authority in the International System', in R.B. Hall and T.J. Biersteker (eds) *The Emergence of Private Authority in Global Governance*, Cambridge: Cambridge University Press.

Hall, R.B. and T.J. Biersteker (eds) (2002b) *The Emergence of Private Authority in Global Governance*, Cambridge: Cambridge University Press.

Hammer, M. and Champy, J. (1993) *Reengineering the Corporation: A Manifesto for Business Revolution*, New York, NY: HarperCollins.

Hampson, F.O., Daudelin, J., Hay, J.B, Reid, H. and Martin, T. (2002) *Madness in the Multitude - Human Security and World Disorder*, Toronto: Oxford University Press.

Harring, S.L. (2000) 'The Diallo Verdict: Another "Tragic Accident" in New York's War on Street Crime?', *Social Justice*, 27(1): 9–18.

Heaton, R. (2000) 'The Prospects for Intelligence-led Policing: Some Historical and Quantitative Considerations', *Policing and Society*, 9: 337–55.

Held, D. (1983) *States and Societies*, New York, NY: New York University Press.

Held, D. (2003) 'Cosmopolitanism: Globalisation Tamed?', *Review of International Studies*, 29: 465–80.

Held, D. (2004) *Global Covenant: The Social Democratic Alternative to the Washington Consensus*, Cambridge: Polity.

Hermer, J., Kempa, M., Shearing, C., Stenning, P. and Wood, J. (2005) 'Policing in Canada in the 21st Century: Directions for Law Reform', in D. Cooley (ed.) *Re-imagining Policing in Canada*, Toronto: University of Toronto Press.

Herrington, V. and Millie, A. (2006) 'Applying Reassurance Policing: Is it "Business as Usual"?, *Policing and Society*, 16(2): 146–63.

Hirschi, T. (1969) *Causes of Delinquency*, Berkeley, CA: University of California Press.

Hobbes, T. (1651/1968) *Leviathan*, Harmondsworth: Penguin.

Hughes, G. (1996) 'Strategies of Multi-Agency Crime Prevention and Community Safety in Contemporary Britain', *Studies on Crime and Crime Prevention*, 5: 221–44.

Hughes, G. (1998) *Understanding Crime Prevention: Social Control, Risk and Late Modernity*, Buckingham, PH: Open University Press.

Human Security Centre (2005) *Human Security Report 2005: War and Peace in the 21st Century*, New York, NY: Oxford University Press.

Innes, M. (2004) 'Reinventing Traditional Reassurance, Neighbourhood Security and Policing', *Criminal Justice*, 4(2): 151–71.

ICHRP (International Council on Human Rights Policy) (2003) *Crime, Public Order and Human Rights*, Versoix: International Council on Human Rights Policy.

Jeffries, F. (1977) *Private Policing: An Examination of In-House Security Operations*, Toronto: Centre of Criminology, University of Toronto.

Joh, E.E. (2005) 'Conceptualizing the Private Police', *Utah Law Review*, 2: 573–617.

John, T. and Maguire, M. (2003) 'Rolling out the National Intelligence Model: Key Challenges', in K. Bullock and N. Tilley (eds) *Crime Reduction and Problem-oriented Policing*, Cullompton: Willan Publishing.

Johnston, L. (1992) *The Rebirth of Private Policing*, London: Routledge.

Johnston, L. (1997) 'Policing Communities of Risk', in P. Francis, P. Davies and V. Jupp (eds) *Policing Futures: The Police, Law Enforcement and the Twenty-first Century*, Houndmills: Macmillan.

Johnston, L. (1999) 'Private Policing in Context', *European Journal on Criminal Policy and Research*, 7: 175–96.

Johnston, L. (2000a) *Policing Britain: Risk, Security and Governance*, Harlow: Longman.

Johnston, L. (2000b) 'Transnational Private Policing: The Impact of Global Commercial Security', in J.W.E. Sheptycki (ed.) *Issues in Transnational Policing*, London: Routledge.

Johnston, L. (2003) 'From "Pluralisation" to "The Police Extended Family": Discourses on the Governance of Community Policing in Britain', *International Journal of the Sociology of Law*, 31: 185–204.

Johnston, L. (2006a) 'Diversifying Police Recruitment? The Deployment of Police Community Support Officers in London', *Howard Journal of Criminal Justice*, 45(4): 388–402.

Johnston, L. (2006b) 'Transnational security governance', in J. Wood and B. Dupont (eds) *Democracy, Society and the Governance of Security*, Cambridge: Cambridge University Press.

Johnston, L. and Shearing, C. (2003) *Governing Security: Explorations in policing and justice*, London: Routledge.

Jones, T. (2003a) 'Accountability in the Era of Pluralized Policing', paper presented at *In Search of Security: An International Conference on Policing and Security*, Montréal, February.

Jones, T. (2003b) 'The Governance and Accountability of Policing', in T. Newburn (ed.) *Handbook of Policing*, Cullompton: Willan Publishing.

Jones, T. and Newburn, T. (1998) *Private Security and Public Policing*, Oxford: Clarendon Press.

Jones, T. and Newburn, T. (eds) (2006) *Plural Policing: A Comparative Perspective*, London: Routledge.

Jordana, J. and Levi-Faur, D. (eds) (2004) *The Politics of Regulation: Institutions and Regulatory Reforms for the Age of Governance*, London: Elgar.

Kaldor, M. (1999) *New and Old Wars: Organized Violence in a Global Era*, Stanford, CA: Stanford University Press.

Kaldor, M. (2003) *Global Civil Society: An Answer to War*, Cambridge: Polity Press.

Karkkainen, B. (2004) 'Post-Sovereign Environmental Governance', *Global Environmental Politics*, 4: 72–96.

Keelty, M. (2006) 'International Networking and Regional Fngagemcnt: An AFP Perspectivc', in J. Fleming and J. Wood (eds) *Fighting Crime Together: The Challenges of Policing and Security Networks*, Sydney: University of New South Wales Press.

Kelling, G.L. (1974) *The Kansas City Preventive Patrol Experiment*, Washington, D.C.: Police Foundation.

Kelling, G.L. (1999) 'Remarks on "Zero Tolerance"', *Criminal Law Rulletin*, 35(4): 374–8.

Kelling, G.L. and Bratton, W.J. (1998) 'Declining Crime Rates: Insiders' Views of the New York City Story', *Journal of Criminal Law and Criminology,* 88(4): 1217–31.
Kelling, G.L. and Coles, C. (1996) *Fixing Broken Windows*, New York, NY: The Free Press.
Kempa, M. (in press) 'Tracing the Diffusion of Policing Governance Models from the British Isles and Back Again: Some Directions for Democratic Reform in Troubled Times', *Police Practice and Research.*
Kempa, M., Carrier, R., Wood, J. and Shearing, C. (1999) 'Reflections on the Evolving Concept of "Private Policing"', *European Journal on Criminal Policy and Research*, 7 (2): 197–223.
Kempa, M., Stenning, P. and Wood, J. (2004) 'Policing Communal Spaces: A Reconfiguration of the "Mass Private Property" Hypothesis', *British Journal of Criminology,* 44: 562–81.
Kerr, P. (2003) *The Evolving Dialectic Between State-Centric and Human-Centric Security,* Canberra: Department of International Relations, Australian National University.
Khagram, S., Clark, W.C. and Firas Raad, D. (2003) 'From the Environment and Human Security to Sustainable Security and Development', *Journal of Human Development,* 4(2): 289–313.
Kitchen, V. (2001) 'From Security to Well-being: A Critical Reflection on the Meanings of Security', *Attaché: An International Affairs Journal at Trinity College and the University of Toronto*, 3: 3–7.
Klockars, C.B. (1985) *The Idea of Police*, Beverly Hills, CA: Sage.

Lange, B. (2003) 'Regulatory Spaces and Interactions: An Introduction', *Social and Legal Studies*, 12(4): 411–23.
Latour, B. (1986) 'The Powers of Association', in J. Law (ed.) *Power, Action and Belief: A New Sociology of Knowledge?* London: Routledge and Kegan Paul.
Latour, B. (1987) *Science in Action*, Cambridge, MA: Harvard University Press.
Lederach, J.P. (1997) *Building Peace: Sustainable Reconciliation in Divided Societies,* Washington, DC: United States Institute of Peace Press.
Leman-Langlois, S. and Shearing, C. (2004) 'Repairing the Future: The South African Truth and Reconciliation Commission at Work', in G. Gillian and J. Pratt (eds) *Crime, Truth and Justice: Official Inquiry, Discourse, Knowledge*, Cullompton: Willan Publishing.
Levi-Faur, D. (2005) 'The Global Diffusion of Regulatory Capitalism', *Annals of the American Academy of Political and Social Science,* 598 (March): 12–32.
Lewis, C. and Wood, J. (2006) 'The Governance of Policing and Security Provision', in J. Fleming and J. Wood (eds) *Fighting Crime Together: The Challenges of Policing and Security Networks*, Sydney: University of New South Wales Press.

Liotta, P.H. (2002) 'Boomerang Effect: The Convergence of National and Human Security', *Security Dialogue*, 33(4): 473–88.

Loader, I. (1997) 'Policing and the Social: Questions of Symbolic Power', *British Journal of Sociology*, 48(1): 1–18.

Loader, I. (1999) 'Consumer Culture and the Commodification of Policing and Security', *Sociology*, 33(2): 373–92.

Loader, I. (2000) 'Plural Policing and Democratic Governance', *Social and Legal Studies*, 9(3): 323–45.

Loader, I. (2002) 'Policing, Securitization and Democratization in Europe', *Criminal Justice*, 2(2): 125–53.

Loader, I. and Walker, N. (2001) 'Policing as a Public Good: Reconstituting the Connections Between Policing and State', *Theoretical Criminology* 5(1): 9–35.

Loader, I. and Walker, N. (2006) 'Necessary Virtues: The Legitimate Place of the State in the Production of Security', in J. Wood and B. Dupont (eds) *Democracy, Society and the Governance of Security*, Cambridge: Cambridge University Press.

Loader, I. and Walker, N. (2006) 'Locating the Public Interest in Transnational Policing', in J.W.E. Sheptycki and A. Goldsmith (eds) *Crafting Global Policing*, Oxford: Hart.

Logan, W.A. (1999) 'Policing in an Intolerant Society', *Criminal Law Bulletin*, July–August: 334–68.

Loughlin, M. and Scott, C. (1997) 'The Regulatory State', in P. Dunleavy, A. Gamble, I. Holliday and G. Peele (eds) *Developments in British Politics*, Houndmills: Macmillan.

Lustgarten, L. (1986) *The Governance of Police*, London: Sweet and Maxwell.

Luzon, J. (1978) 'Corporate Headquarters Private Security' *The Police Chief*, xiv(6): 39–42.

Macaulay, S. (1986) 'Private Government', in L. Lipson and S. Wheeler (eds) *Law and the Social Sciences*, New York: Russell Sage Foundation, pp. 445–518.

Makkai, T. and Braithwaite, J. (1993) 'Praise, Pride and Corporate Compliance', *International Journal of the Sociology of Law*, 21: 73–91.

Makkai, T. and Braithwaite, J. (1994) 'Reintegrative Shaming and Regulatory Compliance', *Criminology*, 32(3): 361–85.

Mandel, R. (2002) *Armies without States: The Privatization of Security*, Boulder, CO: Lynne Rienner.

Manning, P.K. (1977) *Police Work: The Social Organization of Policing*, Cambridge: MIT Press.

Manning, P.K. (1988) 'Community Policing as a Drama of Control', in J. R. Greene and S.D. Mastrofski (eds) *Community Policing: Rhetoric or Reality?* New York, NY: Praeger.

Manning, P.K. (2001) 'Technology's Ways: Information Technology, Crime Analysis and the Rationalizing of Policing', *Criminal Justice*, 1(1): 83–103.

Martin, M.A. (1995) *Urban Policing in Canada: Anatomy of an Aging Craft*, Montreal: McGill-Queen's University Press.
Matin, I. and Hulme, D. (2003) 'Programs for the Poorest: Learning from the IGVGD Program in Bangladesh', *World Development*, 31(3): 647–65.
Matthews, R. (1992) 'Replacing Broken Windows: Crime, Incivilities and Urban Change', in R. Matthews and J. Young (eds) *Issues in Realist Criminology*, London: Sage.
May, P.J. (2002) 'Regulation and Motivations: Hard versus Soft Regulatory Paths', paper presented at the *2002 Annual Meeting of the American Political Science Association*, Boston.
Mayne, J. (1999) *Addressing Attribution Through Contribution Analysis: Using Performance Measures Sensibly*, Ottawa: Office of the Auditor General of Canada.
Mazerolle, L. and Ransley, J. (2005) *Third Party Policing*, Cambridge: Cambridge University Press.
McBarnet, D. (1979) 'Arrest: The Legal Context of Policing', in S. Holdaway (ed.) *Inside the British Police*, London: Edward Arnold.
McBarnet, D. (1984) 'Law and Capital: The Role of Legal Form and Legal Actors' *International Journal of the Sociology of Law*, 12: 231–8.
McBarnet, D. and Whelan, C.J. (1997) 'Creative Compliance and the Defeat of Legal Control: The Magic of the Orphan Subsidiary', in K. Hawkins (ed.) *The Human Face of Law: Essays in Honour of Donald Harris*, Oxford: Clarendon Press.
Mcgeough, P. (2004) 'Private Security Operators Now Make up the Third Largest Armed Force in Iraq', *The Age*, Melbourne, 2 April.
McGinnis, M.D. (1999a) 'Introduction', in M.D. McGinnis (ed.) *Polycentric Governance and Development: Readings from the Workshop in Political Theory and Policy Analysis*, Ann Arbor, MI: University of Michigan Press.
McGinnis, M.D. (ed.) (1999b) *Polycentric Governance and Development: Readings from the Workshop in Political Theory and Policy Analysis*, Ann Arbor, MI: University of Michigan Press.
McLaughlin, E. and Murji, K. (1995) 'The End of Public Policing? Police Reform and "The New Managerialism"', in L. Noaks, M. Maguire and M. Levi (eds) *Contemporary Issues in Criminology*, Cardiff: University of Wales Press.
McLaughlin, E. and Murji, K. (1997) 'The Future Lasts a Long Time: Public Policework and the Managerialist Paradox', in P. Francis, P. Davies and V. Jupp (eds) *Policing Futures: The Police, Law Enforcement and the Twenty-First Century*, London: Macmillan.
Merton, R. (1968) *Social Theory and Social Structure*, New York, NY: Free Press.
Miller, G. (1996) 'Why the Police Department Became a Service', *Blue Line Magazine*, February: 9.
Millie, A. and Herrington, V. (2005) 'Bridging the Gap: Understanding Reassurance Policing', *The Howard Journal*, 44(1): 41–56.

Ministry of Solicitor General and Correctional Services (1991) *Community Policing: Shaping the Future*, Ottawa: Ministry of Solicitor General and Correctional Services.

Ministry of Solicitor General and Correctional Services (1993) *New Race Relations Policy for Police Services Launched*, Ottawa: Ministry of Solicitor General and Correctional Services.

Ministry of Solicitor General and Correctional Services (1996) *Review of Police Services in Ontario: A Framework for Discussion*, Ottawa: Ministry of Solicitor General and Correctional Services.

Monbiot, G. (2003) *Manifesto for a New World Order*, London: The New Press.

Moore, D. and O'Connell, T. (1994) 'Family Conferencing in Wagga Wagga: A Communitarian Model of Justice', in C. Alder and J. Wundersitz (eds) *Family Conferencing and Juvenile Justice*, Canberra: Australian Institute of Criminology.

Mopas, M. and Stenning, P. (2001) 'Tools of the Trade: The Symbolic Power of Private Security: An Exploratory Study', *Policing and Society* 11(2): 67–97.

Møller, B. (2000) *National, Societal and Human Security: General Discussion with a Case Study from the Balkans. What Agenda for Human Security in the Twenty-first Century?*, Paris: UNESCO.

Mulgan, R. (2002) 'Accountability Issues in the New Model of Governance', paper presented at the Political Science Program Seminar, Australian National University.

Murphy, J. (1997) 'The Private Sector and Security: A Bit on BID's', *Security Journal*, 9: 11–13.

Murray, A. and Scott, C. (2002) 'Controlling the New Media: Hybrid Responses to New Forms of Power', *The Modern Law Review*, 65(4): 491–516.

National Commission on the Terrorist Attacks upon the United States (2004) *The 9/11 Commission Report: Final Report of the National Commission on the Terrorist Attacks upon the United States*, New York, NY: W.W. Norton and Co.

National Criminal Intelligence Service (2000) *The National Intelligence Model*, London: NCIS.

Noaks, L. (2000) 'Private Cops on the Block: A Review of the Role of Private Security in Residential Communities', *Policing and Society*, 10: 143–161.

Nygren, A. (1999) 'Local Knowledge in the Environment-Development Discourse: From Dichotomies to Situated Knowledges', *Critique of Anthropology*, 19(3): 267–88.

OECD Development Assistance Committee (DAC) (2003) *A Development Co-operation Lens on Terrorism Prevention: Key Entry Points for Action*, Paris: Organisation for Economic Co-operation and Development Development Assistance Committee (DAC).

O'Malley, P. (1997) 'Policing, Politics and Postmodernity', *Social and Legal Studies*, 6(3): 363–81.
O'Malley, P. (2004) *Risk, Uncertainty and Government*, London: Cavendish.
O'Malley, P. and Palmer, D. (1996) 'Post-Keynesian Policing', *Economy and Society* 25(2): 137–55.
Ontario Provincial Police (1989a) *Annual Report*, Ontario Provincial Police.
Ontario Provincial Police (1989b) *Mission, Philosophy, Plans*, Ontario Provincial Police.
Ontario Provincial Police (1991) *Annual Report*, Ontario Provincial Police
Ontario Provincial Police (1993) *Annual Report*, Ontario Provincial Police.
Ontario Provincial Police (1995a) *Annual Report*, Ontario Provincial Police.
Ontario Provincial Police (1995b) *Focus on the Future: A Process and Model for Change*, Ontario Provincial Police.
Ontario Provincial Police (1996) *Annual report*, Ontario Provincial Police.
Ontario Provincial Police (1997) *How Do We Do It? Manual of Community Policing*, Ontario Provincial Police.
Ontario Provincial Police (2004) *Business Plan 2004*, Ontario Provincial Police.
Osborne, D. and Gaebler, T. (1992) *Reinventing Government: How the Entrepreneurial Spirit is Transforming the Public Sector*, New York, NY: Plume.

Palmer, D. (1997) 'When Tolerance is Zero', *Alternative Law Journal*, 22(5): 232–6.
Paoli, L. (2002) 'The Paradoxes of Organised Crime', *Crime, Law and Social Change*, 37: 51–97.
Parker, C. (2002) *The Open Corporation: Effective Self-regulation and Democracy*, Cambridge: Cambridge University Press.
Parker, C. and Braithwaite, J. (2003) 'Regulation', in P. Cane and M. Tushnet (eds) *Oxford Handbook of Legal Studies*, Oxford: Oxford University Press.
Pasquino, P. (1991) 'Theatrum Politicum: The Genealogy of Capital – Police and the State of Prosperity', in C. Burchell, C. Gordon and P. Miller (eds) *The Foucault Effect: Studies in Governmentality*, London: Harvester Wheatsheaf.
Patten, C. (1999) *A New Beginning for Policing in Northern Ireland: The Report of the Independent Commission on Policing for Northern Ireland*, Belfast: HMSO.
Piper, N. and Hemming, J. (2004) 'Trafficking and Human Security in Southeast Asia – A Sociological Perspective', paper presented at the *Conference on Illegal Migration and Non-Traditional Security: Processes of Securitisation and Desecuritisation in Asia*, Beijing, October.
Porter, T.M. (1995) *Trust in Numbers: The Pursuit of Objectivity in Science and Public Life*, Princeton, NJ: Princeton University Press.
Power, M. (1997) *The Audit Society: Rituals of Verification*, New York, NY: Oxford University Press.
Prenzler, T. and King, M. (2002) 'The Role of Private Investigators and

Commercial Agents in Law Enforcement', *Trends and Issues in Crime and Criminal Justice*, 234: 1–6.

Prenzler, T. and Sarre, R. (1998) 'Regulating Private Security in Australia', *Trends and Issues in Crime and Criminal Justice*, 98: 1–6.

Prenzler, T. and Sarre, R. (1999) 'A Survey of Security Legislation and Regulatory Strategies in Australia', *Security Journal*, 12(3): 7–17.

Raab, J. and Milward, H.B. (2003) 'Dark Networks as Problems', *Journal of Public Administration Research and Theory*, 13(4): 413–39.

Radzinowicz, L. (1948) *A History of English Criminal Law and its Administration from 1750 (Vol. I)*, London: Stevens and Sons Ltd.

Radzinowicz, L. (1956) *A History of English Criminal Law and its Administration from 1750 (Vol. II)*, London: Stevens and Sons Ltd.

Radzinowicz, L. (1968) *A History of English Criminal Law and its Administration from 1750 (Vol. III)*, London: Stevens and Sons.

RAND Corporation (2004) *Collecting the Dots: Problem Formulation and Solution Elements*, Santa Monica, CA: RAND Corporation.

Rankin, K. N. (2001) 'Governing Development: Neoliberalism, Microcredit, and Rational Economic Woman', *Economy and Society*, 30(1): 18–37.

Ratcliffe, J. (2003) 'Intelligence-led Policing', *Trends and Issues in Crime and Criminal Justice*, 248: 1–6.

Ratcliffe, J.H. (ed.) (2004) *Strategic Thinking in Criminal Justice*, Sydney: Federation Press.

Reiner, R. (1992) 'Policing a Postmodern Society', *Modern Law Review*, 55(6): 761–781.

Reiner, R. (1993) 'Police Accountability: Principles, Patterns and Practices', in R. Reiner and S. Spencer (eds) *Accountable Policing: Effectiveness, Empowerment and Equity*, London: Institute for Public Policy Research.

Reiner, R. (1995) 'Counting the Coppers: Antinomies of Accountability in Policing', in P. Stenning (ed.) *Accountability for Criminal Justice: Selected Essays*, Toronto: University of Toronto Press.

Reuss-Ianni, E. and Ianni, R. (1983) 'Street Cops and Management Cops: The Two Cultures of Policing', in M. Punch (ed.) *Control in Police Organizations*, Newbury Park, CA: Sage.

Rigakos, G. (2002) *The New Parapolice: Risk Markets and Commodified Social Control*, Toronto: University of Toronto Press.

Roach Anleu, S., Mazerolle, L. and Presser, L. (2000) 'Third-Party Policing and Insurance: The Case of Market-Based Crime Prevention', *Law and Policy* 22(1): 67–87.

Rose, N. (1996) 'The Death of the Social: Refiguring the Territory of Government', *Economy and Society*, 25(3): 327–56.

Rose, N. (1999) *The Powers of Freedom: Reframing Political Thought*, Cambridge: Cambridge University Press.

Rose, N. and Miller, P. (1992) 'Political Power Beyond the State: Problematics of Government', *British Journal of Sociology*, 43: 173–205.

Rosenbaum, D.P., Lurigio, A.J. and David, R.C. (1998) *The Prevention of Crime: Social and Situational Strategies*, Belmont, CA: West/Wadsworth.

Ruggiero, V. (2003) 'Global Markets and Crime', in M.E. Beare (ed.) *Critical Reflections on Transnational Organized Crime, Money Laundering, and Corruption*, Toronto: University of Toronto Press.

Sampson, R. J. and Raudenbush, S.W. (1999) 'Systematic Social Observation of Public Spaces: A New Look at Disorder in Urban Neighborhoods', *American Journal of Sociology*, 105(3): 603–51.

Scott, C. (2000) 'Accountability in the Regulatory State', *Journal of Law and Society*, 27(1): 38–60.

Scott, C. (2001) 'Analysing Regulatory Space: Fragmented Resources and Institutional Design', *Public Law*, Summer: 329–53.

Scott, C. (2002) 'Private Regulation of the Public Sector: A Neglected Facet of Contemporary Governance', *Journal of Law and Society*, 29(1): 56–76.

Scott, C. (2003) 'Speaking Softly Without Big Sticks: Meta-Regulation and Public Sector Audit', *Law and Policy*, 25(3): 203–19.

Scott, C. (2004) 'Regulation in the Age of Governance: The Rise of the Post-Regulatory State', in J. Jordana and D. Levi-Faur (eds) *The Politics of Regulation: Institutions and Regulatory Reforms for the Age of Governance*, London: Elgar, pp. 145–74.

Scott, C. (2005) *Spontaneous Accountability*, London: Centre for Risk and Regulation, London School of Economics.

SEWA (Self-Employed Women's Association) (2004) *SEWA - About Us*, available online at http://www.sewa.org/(accessed 10 June 2006).

Selin, S. and Chavez, D. (1995) 'Developing a Collaborative Model for Environmental Planning and Management', in *Environmental Management*, 19(2): 189–195.

Shearing, C. (1984) *Dial-a-Cop: A Study of Police Mobilization*, Toronto: Centre of Criminology, University of Toronto.

Shearing, C. (1992) 'The Relationship between Public and Private Policing', in M. Tonry and N. Morris (eds) *Modern Policing*, Chicago, IL: University of Chicago Press, pp. 299–434.

Shearing, C. (1993) 'A Constitutive Conception of Regulation', in P. Grabosky and J. Braithwaite (eds) *Business Regulation and Australia's Future*, Canberra: Australian Institute of Criminology.

Shearing, C. (1996) 'Reinventing Policing: Policing as Governance', in F. Sack, M. Voss, D. Frehsee, A. Funk and H. Reinke (eds) *Privatisierung staatlicher Kontrolle: Befunde, Konzepte, Tendenzen*, Baden-Baden: Nomos Verlagsgesellschaft.

Shearing, C. (1997) 'The Unrecognized Origins of the New Policing: Linkages between Private and Public Policing', in M. Felson and R.V. Clarke (eds) *Business and Crime Prevention*, Monsey, NY: Criminal Justice Press.

Shearing, C. (2000) 'A "New Beginning" for Policing', *Journal of Law and Society*, 27(3): 386–93.

Shearing, C. (2001a) 'Punishment and the Changing Face of Governance', *Punishment and Society*, 3(2): 203–20.

Shearing, C. (2001b) 'Transforming Security: A South African Experiment', in H. Strang and J. Braithwaite (eds) *Restorative Justice and Civil Society*, Cambridge: Cambridge University Press.

Shearing, C. (2006) 'Reflections on the Refusal to Acknowledge Private Governments', in J. Wood and B. Dupont (eds) *Democracy, Society and the Governance of Security*, Cambridge: Cambridge University Press.

Shearing, C. and Ericson, R. (1991) 'Culture as Figurative Action', *British Journal of Sociology*, 42(4): 481–506.

Shearing, C. and Johnston, L. (2005) 'Justice in the Risk Society', *Australian and New Zealand Journal of Criminology*, 38(1): 25–38.

Shearing, C. and Kempa, M. (2000) 'Rethinking the Governance of Security in Transitional Democracies', *Nigerian Law Enforcement Review*, July–September: 33–7.

Shearing, C. and Kempa, M. (2001) 'The Role of Private Security in Transitional Democracies', in M. Shaw (ed.) *Crime and Policing in Transitional Societies*, South African Institute of International Affairs (SAIIA) Seminar Report Number 8, Johannesburg: Konrad Adenauer Stiftung.

Shearing, C. and Leon, J.S. (1977) 'Reconsidering the Police Role: A Challenge to a Challenge of a Popular Conception', *Canadian Journal of Criminology and Corrections*, 4: 331–45.

Shearing, C. and Stenning, P. (1981) 'Modern Private Security: Its Growth and Implications', in M. Tonry and N. Morris (eds) *Crime and Justice: An Annual Review of Research*, Chicago, IL: University of Chicago Press, 3: 193–245.

Shearing, C. and Stenning, P. (1983a) 'Private Security: Its Implications for Social Control', *Social Problems* 30(5): 125–38.

Shearing, C. and Stenning, P. (1983b) 'Snowflakes or Good Pinches? Private Security's Contribution to Modern Policing', in R. Donelan (ed.) *The Maintenance of Order in Society*, Ottawa: Canadian Police College.

Shearing, C. and Stenning, P. (1987) (eds) *Private Policing*, Newbury Park, CA: Sage Publications.

Shearing, C. and Wood, J. (2000) 'Reflections on the Governance of Security: A Normative Inquiry', *Police Practice*, 1(4): 457–76.

Shearing, C. and Wood, J. (2003a) 'Governing Security for Common Goods', *International Journal of the Sociology of Law*, 31(3): 205–25.

Shearing, C. and Wood, J. (2003b) 'Nodal Governance, Democracy and the New "Denizens"', *Journal of Law and Society*, 30(3): 400–19.

Shearing, C., Wood, J. and Font, E. (in press) 'Nodal Governance and Restorative Justice', Regulatory Institutions Network, Australian National University.

Sheptycki, J.W.E. (1995) 'Transnational policing and the makings of a postmodern state', *British Journal of Criminology*, 35(4): 613–35.

Sheptycki, J.W.E. (1998a) 'Review of Schengen Investigated', *International Journal of Evidence and Proof*, 1(4): 246–49.

Sheptycki, J.W.E. (1998b) 'Policing, Postmodernism and Transnationalization', *British Journal of Criminology*, 38(3): 485–503.

Sheptycki, J.W.E. (2002) *In Search of Transnational Policing: Towards a Sociology of Global Policing*, Aldershot: Ashgate.

Sherman, L. (1999) 'Restorative Policing: The Canberra, Australia Experiment', in C. Solé Brito and T. Allan (eds) *Problem Oriented Policing: Crime-Specific Problems, Critical Issues and Making POP Work*, Washington, DC: Police Executive Research Forum.

Simon, J. (1997) 'Governing Through Crime', in L.M. Friedman and G. Fisher (eds) *The Crime Conundrum: Essays in Criminal Justice*, Boulder, CO.: Westview Press.

Singer, P.W. (2001) 'Corporate Warriors: The Rise of the Privatized Military Industry and its Ramifications for International Security', *International Security*, 26(3): 186–220.

Singer, P.W. (2003) *Corporate Warriors: The Rise of the Privatized Military Industry*, Ithaca, NY: Cornell University Press.

Skogan, W. (1990) *Disorder and Decline*, New York, NY: Free Press.

Skogan, W. and Hartnett, S.M. (1997) *Community Policing, Chicago Style*, New York, NY: Oxford University Press.

Smith, A. (1776/1998) *Wealth of Nations*, Oxford: Oxford University Press.

Smith, A. (ed.) (1998) *Intelligence-led Policing: International Perspectives on Policing in the 21st Century*, Lawrenceville, NJ: International Association of Law Enforcement Intelligence Analysts.

South African Homeless People's Federation (2004) *Website for the South African Homeless People's Federation and its Allies, People's Dialogue and Utshani Fund*, available online at http://www.utshani.org.za/(accessed 10 June 2006).

Spitzer, S. (1975) 'Towards a Marxian Theory of Deviance', *Social Problems*, 22(5): 638–51.

Spitzer, S. and Scull, A. (1977) 'Privatisation and Capitalist Development: The Case of the Private Police', in *Social Problems*, 25: 18–29.

Steering Committee for the Review of Government Service Provision (2004) *Review of Government Service Provision*, Canberra: Productivity Commission, Government of Australia.

Steiker, C.S. (1998) 'Supreme Court Review – Foreword: The Limits for the preventive State', *Journal of Criminal Law and Criminology*, 88: 771–808.

Stenning, P. (1995) 'Introduction', in P. Stenning (ed.) *Accountability for Criminal Justice: Selected Essays*, Toronto: University of Toronto Press.

Stenning, P. (2000) 'Powers and Accountability of Private Police', *European Journal on Criminal Policy and Research*, 8: 325–52.

Stenning, P., Shearing, C., Addario, S. and Condon, M. (1990) 'Controlling Interests: Two Conceptions of Order in Regulating a Financial Market', in M.L. Friedland (ed.) *Securing Compliance: Seven Case Studies*, Toronto: University of Toronto Press.

Stinchcombe, A. (1963) 'Institutions of Privacy in the Determination of Police Administrative Practice', *American Journal of Sociology*, 69:150–60.

Stohl, C. and Stohl, M. (2002) 'Networks, Terrorism, and Terrorist Networks', manuscript adapted from a paper presented at the *NCA Organizational Communication Preconference, Communication in Action: The Communicative Constitution of Organization and its Implications for Theory and Practice*, New Orleans, November.

Stohl, C. and Stohl, M. (2004) *Networks and the Bush War on Terror*, Santa Barbara, CA: University of California.

Strang, H., Barnes, G.C., Braithwaite, J. and Sherman, L.W. (1999) *Experiments in Restorative Policing: A Progress Report on the Canberra Reintegrative Shaming Experiments (RISE)*, Canberra: Australian Federal Police and Australian National University.

Task Force on Policing in Ontario (1974) *The Police are the Public and the Public are the Police*, Ottawa: Ontario Provincial Police.

Taylor, R.B. (2001) *Breaking Away from Broken Windows: Baltimore Neighbourhoods and the Nationwide Fight Against, Crime, Grime, Fear and Decline*, Boulder, CO: Westview.

Tilley, N. (2003) 'Community Policing, Problem-oriented Policing and Intelligence-led Policing', in T. Newburn (ed.) *Handbook of Policing*, Cullompton: Willan Publishing.

Toffler, A. (1980) *The Third Wave*, New York, NY: Bantam Books.

Tutu, D. (1999) *No Future Without Forgiveness*, London: Rider.

Uglow, S. (1988) 'The Origins of the Police', in S. Uglow (ed.) *Policing Liberal Society*, Oxford: Oxford University Press.

UNDP (United Nations Development Programme) (1994) *Human Development Report 1994*, New York, NY: Oxford University Press.

Valverde, M. (2001) 'Governing Security, Governing Through Security', in R.J. Daniels, P. Macklem and K. Roach (eds) *The Security of Freedom. Essays on Canada's Anti-Terrorism Bill*, Toronto: University of Toronto Press.

Valverde, M. and Wood, J. (2001) 'In the Name of Security', *University of Toronto Bulletin*, 16: 1.

Victoria Police (2003) *Delivering a Safer Victoria: Business Plan 2003-2004*, Melbourne: Victoria Police.

Victoria Police (2005) *Organised Crime Strategy 2005-2009*, Melbourne: Victoria Police.

Vogelgesang, U. (2003) 'Microfinance in Times of Crisis: The Effects of Competition, Rising Indebtedness, and Economic Crisis on Repayment Behavior', *World Development*, 31(12): 2085–2114.

von Hirsch, A. and Shearing, C. (2000) 'Exclusion from Public Space', in A. von Hirsch, D. Garland and A. Wakefield (eds) *Ethical and Social Perspectives on Situational Crime Prevention*, Oxford: Hart Publishing, pp. 77–96.

Wæver, O. (1995) 'Securitization and desecuritization', in R.D. Lipschutz (ed.) *On Security*, New York, NY: Columbia University Press, pp. 46–86.
Walker, N. (2000) *Plural Policing in a Changing Constitutional Order*, London: Sweet and Maxwell.
Wall, D.S. (in press) 'Policing Cybercrimes: Situating the Public Police in Networks of Security within Cyberspace', *Police Practice and Research*.
Wardlaw, G. and Boughton, J. (2006) 'Intelligence-led Policing: The AFP Approach', in J. Fleming and J. Wood (eds) *Fighting Crime Together: The Challenges of Policing and Security Networks*, Sydney: University of New South Wales Press.
Watchirs, H. (2003) 'AIDS Audit – HIV and Human Rights: An Australian Pilot', *Law and Policy*, 25(3): 245–68.
Weber, L. and Bowling, B. (2004) 'Policing Migration: A Framework for Investigating the Regulation of Global Mobility', *Policing and Society*, 14(3): 195–212.
Weber, M. (1946) 'Politics as a vocation', in H.H. Gerth and C. Wright Mills (eds) *From Max Weber: Essays in Sociology*, Oxford: Oxford University Press.
Weisburd, D., Mastrofski, S.D., McNally, A.M., Greenspan, R. and Willis, J.J. (2003) 'Reforming to Preserve: Compstat and Strategic Problem Solving in American Policing', *Criminology and Public Policy*, 2(3): 421–56.
Williams, J.W. (2005a) 'Governability Matters: The Private Policing of Economic Crime and the Challenge of Democratic Governance', *Policing and Society*, 15(2): 187–211.
Williams, J.W. (2005b) 'Reflections on the Private versus the Public Policing of Economic Crime', *British Journal of Criminology*, 45: 316–339.
Wilson, J. Q. and Kelling, G.L. (1982) 'Broken Windows', *The Atlantic Monthly*, March: 29–37.
Wood, J. (2000) *Reinventing Governance: A Study of Transformations in the Ontario Provincial Police*, Toronto: Centre of Criminology, University of Toronto.
Wood, J. (2004) 'Cultural Change in the Governance of Security', *Policing and Society*, 14(1): 31–48.
Wood, J. (2006) 'Research and Innovation in the Field of Security: A Nodal Governance View', in J. Wood and B. Dupont (eds) *Democracy, Society and the Governance of Security*, Cambridge: Cambridge University Press.
Wood, J. (2006) 'Dark Networks, Bright Networks and the Place of the Police', in J. Fleming and J. Wood (eds) *Fighting Crime Together: The Challenges of Policing and Security Networks*, Sydney: University of New South Wales Press.
Wood, J. and Dupont, B. (eds) (2006) *Democracy, Society and the Governance of Security*, Cambridge: Cambridge University Press.
Wood, J. and Font, E. (in press) 'Crafting the Governance of Security in Argentina: Engaging with Global Trends', in J.W.E. Sheptycki and A. Goldsmith (eds) *Crafting Global Policing*, Oxford: Hart.

Wood, J. and Marks, M. (2006) 'Nexus Governance: Building New Ideas for Security and Justice', in C. Slakmon, M. Rocha Machado and P. Cruz Bottini (eds.) *Novas Direções na Governança da Justiça e da Segurança* Brasilia – D.F.: Ministry of Justice of Brazil, United Nations Development Programme – Brazil, and the School of Law of the Getulio Vargas Foundation – São Paulo, pp. 719–38.
WWF (Working Women's Forum) (2004) *Profile of Working Women's Forum (WWF)*, available online at http://www.workingwomensforum.org/profile.htm/(accessed 10 June 2006).

Young, J. (1998) 'Zero Tolerance: Back to the Future', in J. Marlow and J. Pitts (eds) *Planning Safer Communities*, Dorset: Russell House Publishing.
Yunus, M. (2004) *What is Microcredit*, available online at http://www.grameen-info.org/(accessed 10 June 2006).

Zedner, L. (2003) 'The Concept of Security: An Agenda for Comparative Analysis', *Legal Studies*, 23(1): 153–76.
Zehr, H. (1990) *Changing Lenses*, Scottdale, PA: Herald Press.

Legislation

Police Services Act, R.S.O. 1990, c. P-15.
Adequacy and Effectiveness of Police Services, O. Reg. 3/99 made under the Police Services Act, R.S.O. 1990, c. P-15.

Legal cases

R. *v.* Metropolitan Police Commissioner, ex parte Blackburn, [1968] 1 All E.R. 763, at 769 – *per* Lord Denning, M.R.

Index